Graduate Education in the Chemical Sciences — Issues for the 21st Century

REPORT OF A WORKSHOP

Chemical Sciences Roundtable

Board on Chemical Sciences and Technology

Commission on Physical Sciences, Mathematics, and Applications

National Research Council

NATIONAL ACADEMY PRESS
Washington, D.C.

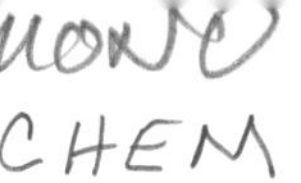

NOTICE: The project that is the subject of this report was approved by the Governing Board of the National Research Council, whose members are drawn from the councils of the National Academy of Sciences, the National Academy of Engineering, and the Institute of Medicine. The members of the workshop organizing committee responsible for the report were chosen for their special competences and with regard for appropriate balance.

Support for this project was provided by the National Science Foundation under Grant No. CHE-9630106, the National Institutes of Health under Contract No. N01-OD-4-2139, and the U.S. Department of Energy under Grant No. DE-FG02-95ER14556. Any opinions, findings, conclusions, or recommendations expressed in this material are those of the authors and do not necessarily reflect the views of the National Science Foundation, the National Institutes of Health, or the U.S. Department of Energy.

International Standard Book Number: 0-309-07130-5

Additional copies of this report are available from:

National Academy Press
2101 Constitution Avenue, NW
Box 285
Washington, DC 20055
800-624-6242
202-334-3313 (in the Washington metropolitan area)
http://www.nap.edu

Board on Chemical Sciences and Technology
NAS 273, National Research Council
2101 Constitution Avenue, NW
Washington, DC 20418
202-334-2156

Printed in the United States of America

THE NATIONAL ACADEMIES

National Academy of Sciences
National Academy of Engineering
Institute of Medicine
National Research Council

The **National Academy of Sciences** is a private, nonprofit, self-perpetuating society of distinguished scholars engaged in scientific and engineering research, dedicated to the furtherance of science and technology and to their use for the general welfare. Upon the authority of the charter granted to it by the Congress in 1863, the Academy has a mandate that requires it to advise the federal government on scientific and technical matters. Dr. Bruce M. Alberts is president of the National Academy of Sciences.

The **National Academy of Engineering** was established in 1964, under the charter of the National Academy of Sciences, as a parallel organization of outstanding engineers. It is autonomous in its administration and in the selection of its members, sharing with the National Academy of Sciences the responsibility for advising the federal government. The National Academy of Engineering also sponsors engineering programs aimed at meeting national needs, encourages education and research, and recognizes the superior achievements of engineers. Dr. William A. Wulf is president of the National Academy of Engineering.

The **Institute of Medicine** was established in 1970 by the National Academy of Sciences to secure the services of eminent members of appropriate professions in the examination of policy matters pertaining to the health of the public. The Institute acts under the responsibility given to the National Academy of Sciences by its congressional charter to be an adviser to the federal government and, upon its own initiative, to identify issues of medical care, research, and education. Dr. Kenneth I. Shine is president of the Institute of Medicine.

The **National Research Council** was organized by the National Academy of Sciences in 1916 to associate the broad community of science and technology with the Academy's purposes of furthering knowledge and advising the federal government. Functioning in accordance with general policies determined by the Academy, the Council has become the principal operating agency of both the National Academy of Sciences and the National Academy of Engineering in providing services to the government, the public, and the scientific and engineering communities. The Council is administered jointly by both Academies and the Institute of Medicine. Dr. Bruce M. Alberts and Dr. William A. Wulf are chairman and vice chairman, respectively, of the National Research Council.

CHEMICAL SCIENCES ROUNDTABLE

Richard C. Alkire, University of Illinois at Urbana-Champaign, *Chair*
Marion C. Thurnauer, Argonne National Laboratory, *Vice Chair*
Alexis T. Bell, University of California, Berkeley
Daryle H. Busch, University of Kansas
Marcetta Y. Darensbourg, Texas A&M University
Michael P. Doyle, Research Corporation
Bruce A. Finlayson, University of Washington
Richard M. Gross, The Dow Chemical Company
Esin Gulari, Wayne State University
L. Louis Hegedus, Elf Atochem North America, Inc.
Andrew Kaldor, Exxon Mobil Corporation
Flint Lewis, American Chemical Society
Robert L. Lichter, The Camille & Henry Dreyfus Foundation, Inc.
Mary L. Mandich, Bell Laboratories, Lucent Technologies
Robert S. Marianelli, Office of Science and Technology Policy
Tobin J. Marks, Northwestern University
Joe J. Mayhew, Chemical Manufacturers Association
William S. Millman, U.S. Department of Energy
Karen W. Morse, Western Washington University
Norine E. Noonan, U.S. Environmental Protection Agency
Janet G. Osteryoung, National Science Foundation
Nancy Parenteau, Organogenesis, Inc.
Gary W. Poehlein, National Science Foundation
Michael E. Rogers, National Institute of General Medical Sciences
Hratch G. Semerjian, National Institute of Standards and Technology
Peter J. Stang, University of Utah
D. Amy Trainor, Zeneca Pharmaceuticals
Jeanette M. Van Emon, U.S. Environmental Protection Agency National Exposure Research Laboratory
Isiah M. Warner, Louisiana State University

Staff

Douglas J. Raber, Director, Board on Chemical Sciences and Technology
Sybil A. Paige, Administrative Associate
Ruth McDiarmid, Senior Program Officer
Susan R. Morrissey, National Research Council Intern

BOARD ON CHEMICAL SCIENCES AND TECHNOLOGY

COMMISSION ON PHYSICAL SCIENCES, MATHEMATICS, AND APPLICATIONS

Preface

The Chemical Sciences Roundtable (CSR) was established in 1997 by the National Research Council (NRC). It provides a science-oriented, apolitical forum for leaders in the chemical sciences to discuss chemically related issues affecting government, industry, and universities. Organized by the NRC's Board on Chemical Sciences and Technology, the CSR aims to strengthen the chemical sciences by fostering communication among the people and organizations—spanning industry, government, universities, and professional associations—involved with the chemical enterprise. The CSR does this primarily by organizing workshops that address issues in chemical science and technology that require national attention.

Graduate education in the chemical sciences was identified by the CSR as an area of broad interest to the chemical sciences community, which has expressed concern about how it should respond to evolving expectations for universities, both in education and research, and to changing patterns in professional employment of advanced degree holders—both in the short and long term. To provide a forum for exploring these concerns, an organizing committee was formed and a workshop was planned for December 1999.

The workshop, "Graduate Education in the Chemical Sciences: Issues for the 21st Century," brought together scientific leaders in government, industry, and academia to explore and discuss the various features of graduate education in chemical science and technology. Using case histories and their individual experiences, speakers examined the current status of graduate education in the chemical sciences, identified problems and opportunities, and discussed possible strategies for improving the system. The discussion was oriented toward the goal of generating graduates who are well prepared to advance the chemical sciences in academia, government, and industry in the next 5 to 10 years.

The papers in this volume are the authors' own versions of their presentations, and the discussion comments were taken from a transcript of the workshop. The workshop did not attempt to establish any conclusions or recommendations about needs and future directions, focusing instead on problems and challenges identified by the speakers. By providing an opportunity for leaders in each of the areas to share their experience and vision, the organizing committee intended that the other workshop participants—as well as readers of this proceedings volume—would be able to identify new and useful ways of improving graduate education and better preparing students for the workforce. We believe that the workshop was successful in meeting this goal.

Workshop Organizing Committee
Rober L. Lichter, *Chair*
Richard C. Alkire
Daryle H. Busch
Thomas F. Edgar
Andrew Kaldor
Janet G. Osteryoung
Michael E. Rogers
Matthew V. Tirrell III
D. Amy Trainor
Francis A. Via
Isiah M. Warner

Acknowledgment of Reviewers

This report has been reviewed by individuals chosen for their diverse perspectives and technical expertise, in accordance with procedures approved by the National Research Council's (NRC's) Report Review Committee. The purpose of this independent review is to provide candid and critical comments that will assist the authors and the NRC in making the published report as sound as possible and to ensure that the report meets institutional standards for objectivity, evidence, and responsiveness to the study charge. The contents of the review comments and draft manuscript remain confidential to protect the integrity of the deliberative process. We wish to thank the following individuals for their participation in the review of this report:

R. Stephen Berry, University of Chicago,
William A. Lester, Jr., University of California, Berkeley,
Lynn F. Schneemeyer, Bell Laboratories, Lucent Technologies, and
John T. Yates, Jr., University of Pittsburgh.

Although the individuals listed above have provided many constructive comments and suggestions, responsibility for the final content of this report rests solely with the authoring group and the NRC.

Contents

Summary

Scholarly research underpins graduate education, but issues in graduate education extend beyond research training. In the chemical sciences, graduate students normally serve as research assistants for faculty, typically with the goal of producing results that can be published jointly by the student and the research advisor. But more often than is the case in other disciplines, graduates in chemistry and chemical engineering are employed by industry—particularly the chemical, pharmaceutical, and biotechnology industries. The unique aspects of the chemical sciences have implications not only for time to degree but also for the larger educational mission of graduate study—including helping students to develop the skills that are likely to be required for multistage career pathways.

The approach to graduate study in the chemical sciences has changed very little in the last 40 years, but the research and educational environment is evolving at a rapid pace. Given the enormity of the economic and human capital investment, it is not surprising that questions arise about the outcomes of the investment. Opinions vary widely about whether graduate education in the chemical sciences needs to change, ranging from an emphasis on not fixing what is not broken to insistence on a complete restructuring. Regardless of one's position on this spectrum, similar questions arise:

- What are the criteria for evaluating the quality of graduate education, and who establishes these criteria?
- For what purpose does the graduate enterprise exist? For what purposes and goals do students choose to seek a graduate degree? Who benefits and what is the product—graduate students, faculty, the research itself, all of the above?
- Must all graduate programs have the same structure?
- Why do faculty choose to have graduate students rather than experienced researchers, especially if research—leading to new knowledge—is the objective?
- To what extent are graduate students educated; to what extent trained?
- How long should—and does—it take graduate students to complete their degrees?

These and other questions are not new and are not unique to the chemical sciences. Indeed, they have captured the attention of the National Research Council, the Association of American Universities, the Council of Graduate Schools, the National Institute for Science Education, the National Science Foundation, a number of private foundations, and of course the array of industries that represent the largest cluster of employers of chemical scientists.

"Graduate Education in the Chemical Sciences: Issues for the 21st Century," a workshop held in 1999 by the National Research Council's Chemical Sciences Roundtable, was organized to address whether and how the various chemical science communities should respond to the types of concerns described above. Speakers were asked to raise questions, rather than to give definitive answers, and to be provocative. The discussion was organized into four sessions that represented a nearly arbitrary framing of the topic: first, a general overview; second, viewpoints on existing circumstances; third, perspectives of and on graduate students; and fourth, some alternative organizational structures. In capturing the presentations and the discussion, these proceedings are intended to broaden the dialog and catalyze mechanisms for participants and others to improve graduate education in or through their own institutions.

OVERVIEW: CHEMICAL SCIENCES IN THE GRADUATE UNIVERSE

Peter M. Eisenberger (Columbia University) opened the session with a presentation that included a review of the 1996 National Science Foundation (NSF) report *Graduate Education and Postdoctoral Training in the Mathematical and Physical Sciences*,[1] which was based on a workshop organized by NSF's Mathematics and Physical Sciences directorate to examine graduate education. After outlining the forces that prompted the NSF workshop and the core findings and recommendations that resulted, he summarized the changes that have taken place within the research and development (R&D) and educational enterprises since the release of that report. The forces that have motivated and are continuing to drive the changes and to shape both graduate education and the overall R&D enterprise in the United States were explored, and emerging trends and institutional and curriculum challenges were discussed.

Dr. Eisenberger argued that we are currently in the middle of a "knowledge revolution" that will have a deep impact on all aspects of our society, but he specifically addressed the emerging impact on R&D and on education. As with the onset of revolutions that have occurred in the past (i.e., the industrial revolution), new challenges will arise that will affect how these enterprises operate. Responding to these changes will have a profound impact not only on graduate education but also on the institutions responsible for education. He stressed the importance of identifying new challenges and addressing them quickly and effectively.

Edel Wasserman (DuPont and the American Chemical Society) accented the need to customize graduate education to match the strengths and weaknesses of the individual student, a task requiring sensitive mentoring. He believes that a diversity of options is desirable in a graduate program. He stated that formal requirements should be targeted to the needs of the good and very good students; the truly outstanding candidate may be a maverick who resists structure. For outstanding students, the university's role is to provide an intellectual and physical environment that can be used for self-education. All students, however, should leave graduate school with the ability to renew themselves continually over a decades-long scientific career.

---

[1]National Science Foundation (NSF), *Graduate Education and Postdoctoral Training in the Mathematical and Physical Sciences*, Report NSF 96-21 (Office of the Assistant, Directorate for Mathematics and Physical Sciences, NSF, 1996). The workshop summary report can be found on the NSF Web site at <www.nsf.gov/mps/workshop.htm>.

The education system must move away from long residencies to complete the Ph.D., Dr. Wasserman argued. Noting anecdotal evidence that residency times greater than 5 to 6 years can lead to a narrower scientific outlook, he observed that the chemical industry requires flexibility in its staff for a rapidly changing work environment, and it needs people who are innovators and who are comfortable working in novel areas. Individuals within the chemical industry need to have the necessary communication skills for interactions between scientists as well as with less technical audiences. Dr. Wasserman pointed out that chemistry has a wide-ranging impact in all fields of science and engineering. Therefore, the chemical sciences need to make it known that chemistry is still a young science, with many open areas to be developed and many exciting opportunities. He concluded by arguing that the educational system must make a transition from what has been largely a single-structure model to a more flexible model, the better to serve both students and the chemical science enterprise.

R. Stephen Berry (University of Chicago) addressed several tensions that exist today in chemistry graduate education: the pull toward interdisciplinarity, the time to achieve a degree, and the intellectual exchange between academia and industry. He questioned whether interdisciplinarity has become too institutionalized: Has institutionalizing interdisciplinarity simply provided a way of making it financially attractive to look beyond traditional bounds rather than increasing the intellectual interactions? He also evaluated the implications of spending an increased amount of time achieving a Ph.D. degree. Vital to resolving the issue of how much time is too much is the need to identify clearly what educational content is required to achieve a Ph.D. Increased interaction between industry and academia is providing new opportunities for research collaborations that factor into the increased time to degree. Care should be taken to ensure that the limiting factor in the time to degree is developing skilled and educated people, not providing to industry the results of substantive research that is enabled by university patents and collaborations with industry. He expressed concern about confidentiality requirements associated with research resulting from industry collaborations. In addition, the problem of impaired information flow has to be addressed before the university loses the capacity to achieve its primary goal of education.

THE CURRENT STATUS OF GRADUATE EDUCATION IN THE CHEMICAL SCIENCES

Lynn W. Jelinski (Louisiana State University) discussed the benefits of interdisciplinary research identifed in her interviews with several graduate students in the chemistry department at Louisiana State University. The case studies arising from the interviews illustrated how interdisciplinary research can increase students' self-confidence, motivation, and ability to compromise. She also found that the faculty members learned how to balance working with both students and external research collaborators. Aspects of external collaborations regarded as important by the student interviewees were compared with aspects the workshop attendees themselves viewed as valuable.

Dr. Jelinski also explored administrative concerns relating to external collaborations to help provide a more complete understanding of external research in the academic setting. She stressed the importance of faculty going beyond what is normally expected of them when they mentor students. Although she pointed out that her study was not highly scientific, she argued that it supports the position that interdisciplinary research is an effective and essential part of graduate education.

Eric G. Jakobsson (University of Illinois at Urbana-Champaign) explored the impacts of information technology on education. He investigated how information technology has blurred the boundaries between research and education as well as between disciplines. He illustrated how the World Wide Web has allowed movement across boundaries of a discipline, using as an example the National Computational Science Alliance Information Workbench, a Web-based program that performs many different functions but is designed to look like a single program. In particular, he explored the capabilities of the

Biology Workbench and showed how its users are able to search for various protein structures, find literature references, view a protein's tertiary structure, find sequence homogeneity with similar proteins by using the BLAST program, and much more. The seamless access to various tools illustrated in his presentation is a feature that makes these programs attractive to biologists as well as those from other related disciplines such as chemistry.

Angelica M. Stacy (University of California, Berkeley) examined how various teaching models affect student learning. In the first model the teacher transmits knowledge to students, who receive and memorize it. In this situation the student does not typically develop a working understanding of the material. An alternative constructivist model features the teacher as a guide to help students work through chemical problems and develop a solution that becomes part of their understanding. Results of an experiment designed to explore which model was more effective indicated that the students taught with the constructivist model performed better at the end of the semester.

Dr. Stacy emphasized the need for improved training for new teaching assistants (TAs), who find themselves responsible for teaching college lectures and/or laboratory sections without any prior teaching experience. She proposed that a solution to this common problem was to employ supplemental instruction programs to equip new TAs to handle their new positions better. In all cases, the faculty's role is to support and assist their students in the learning process.

KEEPING AN EYE TO THE FUTURE IN DESIGNING GRADUATE PROGRAMS

Marye Anne Fox (North Carolina State University) discussed various issues that affect graduate education. These ranged from the effects of advances in information technology on how science is conducted to the special obligations of research scientists to K-12 education as a result of a "covenant with the nation." To help explore improvements that can be made to enhance graduate education, she also discussed topics related to funding, such as reasons for federal government support of basic research and alternative ways in which graduate students could be supported.

Dr. Fox called for the integration of research into education and urged scientists to educate the general public about the purposes of their research. She charged faculty members to evaluate their students for their readiness to work outside their own specializations. She also encouraged the faculty to take undergraduate students into their research and called for research scientists to establish ties with and offer assistance to colleagues who are involved in teacher education. Finally, she encouraged increased interactions between faculty members and local industry colleagues, including investigating the possibility of co-sponsoring student research projects.

THE GRADUATE STUDENT PERSPECTIVE

Karen E. Phillips (Columbia University), an advanced graduate student, discussed the Columbia Chemistry Career Committee (C^4) that she initiated to help prepare graduate students for the transition to professional employment. Her motivation for developing C^4 was the observation that graduate schools, although they generally provide a model for students with career paths leading toward academic research universities or careers in industry, still largely ignore the needs of future small-college or community-college instructors. The C^4 activities allow students to gain a better understanding of life after graduate school by participating in an interview workshop, a résumé workshop, a trip to an industrial plant research facility, and a panel discussion group. She elaborated on the panel discussion and explained how it investigated general job-market issues such as funding. The events had a positive effect on the students planning them as well as on those attending. C^4 is currently making plans to hold another panel discussion that will focus on career options for teachers with graduate degrees.

Jonathan L. Bundy (University of Maryland, College Park) addressed the issue of undergraduate education from his perspective as a relatively new graduate student. He viewed reforming undergraduate education as the most important contribution to improving graduate education in chemistry. The three areas that deserve attention are undergraduate-level research experiences, advising and mentoring, and ensuring that students have an adequate foundation in chemical fundamentals. He also argued that graduate programs can help improve undergraduate education by instituting teaching fellow or faculty apprentice programs to enhance the teaching assistantship experience for those who ultimately may pursue a career as a faculty member at an undergraduate institution. Maintaining the "raw material" is essential to continuing the excellence of our graduate institutions.

Judson L. Haynes III (Procter & Gamble), a recent graduate, discussed his experiences as both an undergraduate and a graduate minority student in chemistry. He likened his graduate education to a modern-day version of the mythical labyrinth of Crete, complete with minotaurs that tried to block his progress through the maze of higher education. As background for his graduate experiences, he discussed his undergraduate education, including his participation in the MARC (Minority Access to Research Careers) program. This federally funded program provided, among other benefits, funding for students to conduct research and to visit universities with different programs, opportunities that helped him to be better prepared for graduate school. After describing some of the issues associated with being a minority student and the important skills he developed during his graduate career, he concluded by emphasizing the importance of getting students involved in chemistry as much and as early as possible and keeping them motivated.

Richard A. Weibl (Association of American Colleges and Universities) provided a summary of the Preparing Future Faculty (PFF) program, sponsored by the Council of Graduate Schools and the Association of American Colleges and Universities, with financial support from the Pew Charitable Trusts and the National Science Foundation. The PFF program is designed to give graduate students who aspire to be college or university faculty a better understanding of the teaching, research, and service roles of faculty members. He discussed the assumptions, concepts, and activities that define the program and the role of PFF in the participating chemistry departments.

ALTERNATIVE STRUCTURES AND ATTITUDES

François M.M. Morel (Princeton University) examined an experimental program started in 1997 at Princeton University and designed to broaden graduate education in environment-related areas by introducing a policy component into the Ph.D. requirements. The Princeton Environmental Institute-Science, Technology, and Environmental Policy (PEI-STEP) program has received support from the Chemistry Division of the National Science Foundation and is administered by PEI. He discussed the requirements of the program and outlined where students currently enrolled are likely to go following completion of their degrees. He pointed out that negative perceptions of such programs, which can stem from an attitude on the part of some faculty that students in these types of programs as not serious about scientific work, can hinder greater interest in such programs.

J. Michael White (University of Texas at Austin) examined the need for chemists to adapt and reinvent themselves to meet the changing demands of the workforce. He discussed his experiences as director of the National Science Foundation-supported Science and Technology Center for the Synthesis, Growth, and Analysis of Electronic Materials. He explained how this interdisciplinary program has integrated graduate research through collaborations between faculty in electrical engineering, chemistry, physics, and chemical engineering. The interdisciplinary nature of this program benefits students throughout their life by providing them with a broad knowledge base. Specific issues arising in the program were also discussed. He emphasized the need for seamless, high-quality education, investments in high school education, and a commitment to lifelong learning.

Ronald T. Borchardt (University of Kansas) focused on how graduate education in the Department of Pharmaceutical Chemistry at the University of Kansas has evolved over the past 30 years and described how a National Institutes of Health predoctoral training grant was used to bridge the gap between traditional disciplines. In many cases, reduction of the gap has been driven by the biotechnology revolution, which demands interactions between various disciplines (e.g., chemistry and biology or pharmacy and engineering). His historical overview of the University of Kansas program covered three periods: prerevolution, postrevolution, and the future. He underscored the changes introduced to generate Ph.D. scientists who could compete effectively for jobs in the emerging biotechnology industry.

Dr. Borchardt also discussed the nonprofit Globalization of Pharmaceutics Education Network, Inc., a program started in 1996 to increase graduate students' interactions with students and faculty from universities in other countries. Throughout, he stressed the importance of developing effective strategies oriented toward future needs.

1

The Challenges to American Graduate Education

Peter M. Eisenberger
Columbia University

I am going to address the challenges to American graduate education by first reviewing the results of a workshop held in 1995 by the Mathematics and Physical Sciences (MPS) Directorate of the National Science Foundation (NSF) on graduate education. After reviewing those findings, I will try to update the environment in which that workshop was carried out to the present and highlight some of the changes that have occurred, even in the short period of time since then. As you will see, most of the core conclusions that came from that workshop are still relevant today.

DRIVING FORCES

Before reviewing the findings of the 1995 MPS workshop, I would like to review the driving forces that were in place at that time and prompted holding the workshop and how these forces have evolved to the present (Box 1.1). These forces affect the research and development enterprise and therefore the educational enterprise that is critical to it.

The first force is the almost complete transition to a civilian focus, except for the recent security flap at the Department of Energy (i.e., the allegations of Chinese espionage). This is going to cause reverberations in that organization for some time. Another force present in 1995 was concern about competition from the Japanese. I think our country is more self-confident now about its ability to perform in a global marketplace. That is a major shift since that time frame. The final force I would like to mention is the heightened concern about America's increasing global impact.

I would also like to touch on an emerging trend. We are entering a new age in which the topics are going to be new applications of knowledge to many new problems. This "knowledge revolution" will be very good for us in the sense that knowledge will be recognized as important.

On the other hand, the knowledge revolution is going to get us involved with many more messy problems, and it will emphasize our obligation—all of us, as an enterprise—to communicate with the society we are a part of. I will return to the changing emphasis later, when you will see how it dovetails with what was anticipated in the 1995 workshop and how it has become even more important today.

BOX 1.1
Major Factors Driving Change in Research and Development

- Replacement of defense objectives by civilian and commercial needs for research and development
- Global competition and growing concern about America's global impact
- The fast pace of innovation
- The information age, the Internet, and changing organization and operational practices
- The increased complexity of important scientific problems, emerging technologies, and societal problems
- The growing understanding between science and technology generators and decision makers

All of these factors have an impact on graduate education in particular and on the overall educational framework as well.

RESULTS OF THE 1995 NSF WORKSHOP

About 100 people, including students, participated in the day-and-a-half meeting at the NSF. They came up with a series of findings, which led to a series of recommendations.[1]

The first finding relates to funding. In 1995 we were concerned about facing a large budget deficit and about the little discretionary money available because of the combined expenses of national security and social programs. This constraint on funding was a major concern at that time. It still exists even in today's economic boom in the sense that we have to compete with other entities for the public dollar, and we can no longer depend on our role in national security to give us a preemptive right to resources. Now we have to compete for available funds and demonstrate our value.

While federal research dollars have not declined, the shift away from military spending has caused significant dislocations in many institutions. We are in a process of completing our adjustments to the shift in spending priorities. Because of this transition, the workshop participants concluded that we should develop programs that address social priorities. In other words, if military expenditures were going to go down, what would go up? The needs identified were social concerns, concerns associated with the economy, and concerns associated with the quality of life. If we wanted to develop and apply for new resources in those directions, it would be important for us to demonstrate the value to society of the academic research enterprise and provide concrete benefits, similar to what had been done in the national security area.

The second finding was supported heavily by the industrial participants. They observed that, by and large, the students coming from graduate programs were not well prepared to contribute within an industrial setting. On one hand, our educational enterprise is the envy of the world. Students from around the world come to study here. Yet, the participants at that workshop believed strongly that

[1]National Science Foundation Workshop Report: *Graduate Education and Postdoctoral Training in the Mathematical and Physical Sciences*, Report NSF 96-21 (Office of the Assistant Directorate for Mathematics and Physical Sciences, NSF, 1996). The workshop summary report can be found on the NSF Web site at <http://www.nsf.gov/mps/workshop.htm>.

graduate education in the physical sciences and mathematics was admirable only from one perspective, that of training someone to go back into academia, while the majority of graduates were going into nonacademic careers. The fact that students were so narrowly trained was viewed as healthy neither for them nor for the overall enterprise.

Science moves forward, and individual fields are becoming more specialized. One result is that students will have to develop skills beyond their scientific expertise. They will have to learn communication, management, and other interpersonal skills because they are going to be asked increasingly to work in teams, both within and outside academia.

The third finding is something that I think is familiar to all of us. As the American university enterprise becomes increasingly competitive, the emphasis on research becomes even more exaggerated. This is having a profound impact on the relationship of the faculty to the students and on the educational experience of the students. A strong bias exists toward research, even at schools that stress education. At schools of this type, such as Princeton University, the faculty still spend a minority of their time on teaching and are largely focused on research. That has an impact that shapes—some would say distorts—the nature of the educational experience.

One distortion is felt in how the strong emphasis on research contributes to the long time that it takes students to get Ph.D.s. Some statistics were presented that indicated that the length of time was getting longer, not shorter. It was speculated that the time had to do with the professors' unwillingness to let successful students go because they were essential to the professors' successes. They had to help the professor write the next grant application or get a paper out to satisfy the previous grant. The students' time was better correlated with that than with the actual requirements of their education. The students were complacent about this. They did not particularly want to go into the real world, because they were having a good time, in many cases, in the educational framework itself. As I will say when we get to the recommendations, the workshop participants believed that society was thereby the loser.

The education enterprise at that time was viewed as a training ground for people going back into academia. It did not value the professional master's programs in the sciences, such as the chemical sciences, which prepared students to go directly to industry but did not satisfy the academic requirements that were believed necessary to stay in academia. Those people obtaining a master's degree were viewed as second-class citizens.

The fourth finding is a very important issue. For some reason, academia seems to be one of the more resistant institutions in America to include diversity. This has become increasingly clear as evidenced by the limited involvement of women and minorities in the faculty ranks in the universities in particular. In mathematics and the physical sciences, the disciplines of physics and mathematics have been the most resistant to change.

As has been documented and recently published in a study at the Massachusetts Institute of Technology (MIT),[2] this problem seems to have taken on a new form. There are underrepresented groups in the pipeline, but, as these individuals move up to higher levels, the resistance to them appears to become greater. It is the academic equivalent of the glass ceiling.

Much discussion at the 1995 workshop focused on the decline in the number of U.S. students. Some believed there was nothing to worry about, since this country has always obtained talent from abroad. Others believed very strongly that it was important to have a good representation of U.S. students.

A fifth finding addressed the time needed to complete a Ph.D. degree. The length of time to the Ph.D. became a dynamic issue in the workshop. It seemed to be the distortion in the education process that symbolized the concerns of the workshop participants. The participants could understand that if

[2] "A Study on the Status of Women Faculty in Science at MIT," *The MIT Faculty Newsletter,* vol. XI, no. 4, March 1999.

students were going to return to an academic career, or never leave academia, it did not matter when the demarcation points between graduate student, postdoctoral fellow, and assistant professor occurred.

If you were going to leave academia, however, it would make a difference. People who spend a long time getting their Ph.D.s are at a disadvantage compared with their peers in other disciplines and skilled professions who enter industry earlier and get a big jump in their careers. In many cases, the high performers in industry are identified early. What you learn is that in industry how fast you move is exceptionally important. Spending many years in a Ph.D. program is not good for your success.

Another issue is that the federal government and society are supporting graduate education because they want the students to contribute to society. To support students who spend more time in the research enterprise is more costly and may be unfair to taxpayers.

To address the issue of broadening students' skills, the minors were introduced. The first objective was to give students additional scientific breadth. For example, if you are a chemist, you would take some biology, a little physics, or maybe even some engineering to broaden your knowledge base. The second motivation was to develop other skills that might be useful to students who are not going into academia. A program could be developed offering them a minor in business, law, or engineering that would give them, in addition to their main expertise, other skills that could be useful to them.

The objective was to develop in the physical sciences something that already had been done in other areas at the master's level. The idea was to make master's-level programs directly dovetail with employment—just as in business or law school. It does not, after all, do any good to have a master's program if the students cannot get jobs. So, we were encouraged in the workshop to try to work with industry and develop specific master's programs that would produce people that industry or society in general would be more interested in employing. At least in my institution, I see significant progress being made on that front.

The next recommendation was another hot button that was discussed many times. It is very important to help the students coming out of their graduate careers to be better at communicating interpersonally, in presentations, and in writing. From many different perspectives, this is a major deficit that needs to be corrected.

The shortening of the time to Ph.D. was a strong focus that attracted much discussion. Many workshop participants supported a fixed time limit, such as that for financial support at Princeton. The NSF workshop participants decided, instead, to recommend what I will call knowledge feedback by distributing best practices and making people aware of the average time in each discipline to get a Ph.D. If an institution or an individual were to fall well outside the norm, best practices would provide the basis for comment by a program officer or reviewer and might offer a way to address this issue. In the end, no definitive recommendations were made for shortening the time to degree.

For broadening skills, the workshop recommended the use of internships. A program has come into being that has created opportunities for alliances with industry. The idea is to develop joint programs in which students would work on a project that had simultaneous industrial and academic support. There was also an idea for joint funding to create new programs. One program that I am aware of that resulted from the workshop was a collaboration between Lucent and the NSF. Again, the idea was to give students some experience outside the academic walls to broaden their perspectives.

The effort to address the education-research balance structurally was very important for the NSF workshop. There is a basic problem with the principal investigator (PI) and graduate student relationship. Giving the money to the PIs makes their motivations dominate the educational enterprise, while those of the student and the institution are less important. Two ideas were proposed to get a better balance. The first was to give the money to the students, who would then shop around and pick the people with whom they want to work. Professors who do not have a good reputation or who require an excessively long time to degree are not going to get many students. Also, this will broaden the selection of interesting subjects to pursue.

The second idea was to give more resources to institutions, or to have new entities within universities that bring groups together in collective enterprises to get support. The objective was to shift the responsibility of education to the larger whole and not have it be the total responsibility of an individual acting one on one with the students.

Another idea was to use a similar approach with respect to the nagging problem of underrepresented groups. The problem would have more focus by requiring proposals to include bringing people together as one of the objectives. One of the criteria by which a proposal would be judged would be its ability to improve diversity.

NSF also introduced the Integrative Graduate Education and Research Traineeship (IGERT) program, which by now has had three national competitions. I don't know whether its generation is related directly to the MPS workshop, but it is so much on target in addressing all the issues identified in that workshop that I assume it was. Basically, IGERT's criteria are very different from those of previous programs. It is intended to support comprehensive, multidisciplinary research projects, so it is explicitly trying to broaden the coterie of people to whom the students will be exposed.

IGERT grants arrange for training activities in some of the "soft" skills, such as understanding ethics, improving group interactions, and teaching people how to work together in groups. In a more traditional vein, exposure to state-of-the-art research tools is also a criterion. Increasing the number of people from underrepresented groups is a criterion for assessment of performance.

The awards, which range up to $500,000, are not for research but for supporting students. Professors must have their own support for research. I think the research university community accepted the program enthusiastically. In the first competition in 1998, 620 preliminary proposals were submitted. Of these, 65 full proposals were invited, and 17 were selected. At my institution the program promoted discussions and dialogues across boundaries that had not occurred before. It has resulted in creative proposals across those boundaries.

The particular focus of IGERT is consistent with the point of view I have tried to express here today, in that it is trying to increase crossing the boundaries between the natural sciences and the social sciences. That is one of its characteristic features. There are additional examples of NSF programs that I consider as emblematic of this goal. One involves chemists working together with ecologists to study pollution in lakes. Besides talking with ecologists and other scientists who work on pollution problems, it has allowed students to become involved with regulators at both the state and federal level. Another program provided an opportunity for chemists to work together to design new molecules and then collaborate with their engineering colleagues to incorporate the molecules into devices. At the same time, they worked with industry.

The preceding examples illustrate how to broaden the exposure of students. In many cases, an effort was made to avoid lengthening the time to Ph.D. Such programs should be consistent with the overall educational objectives, although some flexibility may be needed in the requirements for the major to allow the students to pursue these interests while getting their Ph.D.

UPDATING THE WORKSHOP—CHANGES SINCE 1995

Funding has stayed stronger than we expected in 1995 for two reasons. One is the booming economy. Also, in 1995, a debate was going on between the Democrats and the Republicans about the appropriate role of the federal government in supporting research. I think in reality they both agreed, but political factors forced them to disagree. Now, they have decided they do agree and that it is important for the nation's health to provide strong support for research. I believe that everybody recognizes that the so-called knowledge revolution or knowledge age is in a transition that gives America a competitive advantage. We have the social, economic, and intellectual infrastructure to make America highly competitive in a global economy that is based on expertise and knowledge.

Since 1995, it has become clear to all of us that the Internet is emerging as a major force. It is beginning to penetrate in many forms the academic and intellectual enterprise. We are just at the beginning of that and don't know yet how far-reaching the changes will be. This suggests a manufacturing analogy. I think the university of today is going to look like a small textile plant in New England before mass production and technology.

Because of the changing ways that research is supported, we are going to have a continued shift to what I call messy problems. We can see it in many ways. The National Science Board has considered the environment as a new program for the NSF. Studies of urban issues are growing at many schools as well as new multidisciplinary neuroscience centers. We are beginning to see this shifting in many places to problems more directly related to societal needs.

As part of this shift—not surprisingly, to my way of thinking—the whole question of the postmodern debate arises. The reason is that science has left the sanctity and security of working on basic science and having military impact. In a complicated way, we are becoming increasingly involved in societal questions that are not necessarily amenable to the simple reductionist answers of yes and no.

For example, is global warming occurring or not? This is not a simple question, and there is no simple answer. Scientists disagree. It has made the entire process by which the enterprise operates more problematic. What do you do when there is ambiguity? What do you do when what you choose to focus on makes a difference in the social context? What you choose can shape the social agenda of a country. That makes the debate extremely important. In that sense, the postmodernists have a point.

However, the fact that the apple will fall to the ground—if it falls from the tree—is not a subject for debate. There is a middle ground that those of us on the reductionist side would be ill-advised to ignore. I have said several times that we are undergoing a transition from an industrial to a knowledge age, not an information age. There is a big difference. You can have all the information you want, but it may not do you any good. Knowledge, however, will have an impact.

The good news is that we are in a growth area. The demand for educated and trained individuals will continue to increase. It will pervade all aspects of life, and it will be very important that our students obtain a good education.

Because we academics are at the heart of this transition, we are going to be subjected to significant changes. There is a real possibility that new institutions will replace the existing ones. History shows that it is rare to go through a transition as profound as this and to have the original institutions emerge unchanged or even continue to exist. The changes are just beginning, and they are going to happen very quickly.

The transition from military to social objectives, and the transition of focus to issues such as global warming and urban problems, will have us continually interacting with the decision makers. It is going to be very complicated and very messy, because we are going to have experts on both sides of the debates. We are going to be worried, and rightfully so, about how respect for our professions can be maintained if the experts can widely disagree about something that is supposed to be understandable.

We are going to have to get used to the idea of giving advice without giving definitive answers. Instead, we will have to give a portfolio of options with probabilities of their outcomes. Can you imagine Congress dealing with that today? That is what must happen. We are going to have to evolve a much more sophisticated way of communicating our knowledge to decision makers. Otherwise, the feedback to us is going to be awful. We will get blamed for all the mistakes. On the other hand, if we can solve the problems, we would be eminently useful.

The institutional impact of these changes is that we will see Internet-based alternatives to our offerings. I believe that people will try to hire some of our best faculty, who are not fully employed by their universities. They are sometimes employed only nine months a year and four days a week. The alternative institution is going to take the best faculty and put them on the Internet in some form to deliver high-quality education to a large number of people. The new institutions will be able to afford to pay these faculty well.

It is just a matter of time before a credible version of this occurs. New institutions, being more innovative, and not restricted by faculty meetings, are going to be more responsive than universities to these emerging opportunities. I don't know if you noticed in recent news reports that MIT and the University of Cambridge have come up with a new partnership.[3] I think we are going to see more schools getting together to try to strengthen themselves in the face of these changes.

We can try to make universities more adaptive by crossing disciplinary boundaries. We are going to have to think of how to return a little more control to the center, so that the institution can direct itself. We need to avoid the "tragedy of the commons"[4] in the university structure.

We need to regard students more as customers who have many choices. If we don't, the students will go elsewhere. Some universities, for example, are thinking about developing their own Internet alternatives to the traditional university environment.

These changes make the 1995 MPS workshop recommendations even more important. I think that the NSF workshop identified many of the core issues. I don't know if the workshop participants successfully addressed all of the issues or if you agree with their recommendations. I hope to see that clarified in the discussion. However, I think they were on target. The urgency to address the issues has increased, and the outlook for science is much stronger than it was in 1995. An important part of that positive outlook is the need for us to adapt. If we do not adapt and respond to the challenges, somebody else will do it for us.

DISCUSSION

Eric Jakobsson, University of Illinois at Urbana-Champaign: We don't need to wait for credible proof that you can do high-quality graduate programs on the Internet. Our graduate school of library and information sciences at the University of Illinois offers an Internet distance master's degree. The only time required on campus is a couple of weeks at the beginning of the program and a short time at the end. In the latest *U.S. News and World Report* rankings of graduate programs, this program is tied for first in the nation.[5] It is absolutely possible, and it is being done, to offer a nationally ranked graduate program on the Internet.

Peter Eisenberger: Thank you for that comment. At Columbia also, our engineering school has developed a program that is offered on the Internet.

Isiah Warner, Louisiana State University: I sometimes like to play the devil's advocate, and I am going to do that now. When we talk about training graduate students for industry, we encounter the problem that industry does not have one model, whereas academia has a single model. It is easier to

[3]"International Collaboration: University of Cambridge to Team Up with MIT," Michael Hagmann, *Science 1999,* November 12, 286: 1271.

[4]Garrett Hardin. The tragedy of the commons. *Science*, 162:1243-1248, 1968.

[5] "America's Best Graduate Schools," *U.S. News & World Report*, August 30, 1999.

train students for a single model. What I am asking is that when you start to train students in a particular focus area of industry—let's say to train them for the microelectronics industry—this is training them for a very narrow area. How do you overcome that problem and still train people for industry? I have had students who say, "I was well trained for industry," and others who say, "I was not well trained for industry," because of its diverse operations.

Peter Eisenberger: I believe that education should give people skills, not facts. Because it has been connected to the research enterprise in graduate education, it has stressed facts. What I am suggesting is that we focus on flexible skills. People in my son's generation are going to have seven or eight jobs in their lifetime, unlike those in my generation, who were likely to have only one.

Isiah Warner: That is the old model. Even if we don't always keep it.

Peter Eisenberger: The old model was to give them facts, not skills.

Laren Tolbert, Georgia Institute of Technology: I found much to agree with in what Dr. Eisenberger said today, but in the spirit of this meeting, I also found much to disagree with, so let me emphasize the latter. We have heard over a number of years that what we need in graduate education is something similar to the Saturn assembly line and that the education process should work faster and do more in a shorter period of time. I think that has been inflicted on us, to a large extent, by success in manufacturing, but educating people is different.

One of the issues of disagreement is exactly the issue that just came up: What is graduate education all about? It is not about knowledge and not about facts but rather it is about thinking. I tell my students that when they leave I expect them to be able to enter a bare room, be told to work on surfactants, for example, and make something happen. That is the skill I think our graduate program is famous for in the United States. I think we need to be careful about adding more expectations onto our graduate students and trying to do it in 4 years. No consensus exists regarding the contents of graduate education. I think that is really the key issue: What is it that we are expecting a student to be like when she or he gets a Ph.D.? Although we have not yet resolved this issue, this workshop is a step in that direction.

Peter Eisenberger: I couldn't agree more about what you are trying to do in education. I would like to make some comments, though. Rarely would a student go into a room alone. More and more people are working on a team, and they are going to be working with things that may not be surfactants or may not be chemistry but may be things related to chemistry. Therefore, education now for the majority of students is not going to focus on a particular area such as surfactants, because that is not what they are going to do when they get out. If you want to orient education programs to the majority of students, you cannot argue with what was said. The programs should therefore be oriented to cover both situations.

Stanley Pine, California State University, Los Angeles: As chair of the American Chemical Society (ACS) Task Force on Graduate Education, I want to be sure all of you are aware that, as of January, the American Chemical Society will have an office of graduate education. Through an invitational conference earlier this year, we decided that it was time to really focus, at least in the chemistry area and chemical sciences, on activities in graduate education. Clearly, ACS has been doing a lot, but it has been scattered. As a professional society, looking at graduate education seemed like a timely thing to do.

One of the emphases will be looking at consensus. Is there a right way? There is probably not one way to do graduate education, so how can all of this be brought together. I am hoping, through this conference, that I will have many opportunities to talk to you, as will other members of that task force, to see how ACS can play the kind of broad leadership role that I think it has for a long time. We think this is part of the continuum of education. It isn't just the graduate level. It starts at pre-kindergarten and continues through K-12 and undergraduate education. ACS has all of those areas well covered, and now we hope that this graduate education office will effectively extend that continuum.

Peter Eisenberger: The other comment I would make is that I think our field is getting a little bit more like medicine. The internship and the initial years of the job are also an important part of training. The boundaries have become blurred between having to learn more within the walls of academia and what you can learn on the job. I think many people in industry observed that there was a retraining process that had to go on.

Lynda Jordan, North Carolina Agricultural and Technical State University: I would like to address the issue of increasing the number of women and minorities in the chemical sciences. In fact, a major issue is the attrition of minorities and women graduate students from the graduate education process in this country. As to the former meetings that were held, let us start by examining who was invited to participate in those meetings and were those participants representative of all the various types of graduate education programs in this country? Without proper representation, how could the appropriate pertinent issues be brought to the table or addressed? Who contributed the information that examined what the issues are in this area?

The second point I would like to address is directly focused on increasing the number of minority graduates in the chemical sciences. A lot of emphasis, at least during the last 10 years, has been on the recruitment of minority graduate students in sciences and engineering. We can see from our data that recruitment is not the major issue. The retention and successful matriculation of minority graduate students in the sciences and engineering are the major points of concern. There is no question of the intellectual capability of the students or the students' abilities to do the work. What is lacking and is not in place is the infrastructure for supporting minority graduate students once they get to the university system. How amenable are we, as educators and administrators, to addressing our own intellectual and cultural comfort zones? Are we able to accept graduate students who are different from us into our education arena, at our universities, and really educate them without the obstacle of our bias? This problem exists in the chemical sciences, not only in the education arena, but also in the workforce as well. How well are we equipped, or willing to surpass our own personal limitations of diversity, to deal with these "soft education" issues if we really want to change the demographics of the chemical science professions? At the eve of the new millennium we have to address diversity effectively, or we will have a greater problem than we do now.

Peter Eisenberger: I have addressed this many times to no avail in my own institution. There has to be cultural change within these institutions, and they are not willing to take it on. They are not willing to ask the professoriate to confront their biases. I am very sympathetic personally to the concept, but until they do that there will be no change. When you look at the other parts of our society that have made those changes, they have confronted their workers directly. In universities, they are unwilling to do that.

Peter K. Dorhout, Colorado State University: The IGERT program and a number of the other issues that you raised represent a number of creative ideas for making changes to graduate programs. I want to point out an aspect for caution, though, with respect to evaluating our faculty in terms of tenure and promotion. I have been in the smoke-filled room a number of times, and I have not seen much change in how we are evaluating young faculty. The young, energetic faculty are the ones that we hope will make an impact in terms of these changes that you are suggesting.

Giving fellowships that would grant to the individual student and not to the faculty member is going to have an impact on how that faculty member could be viewed in the future. I don't see changes in the letters that are being written for faculty members. So, I think the onus is on all of us to try to improve the way in which we write tenure letters.

It is still being said in the smoke-filled room that small grants are not real grants, or that a person does not have enough money even though they have 12 graduate students and 4 of them have been granted Ph.D.s. There is still the idea that only a $300,000 NSF grant or a $600,000 NIH grant counts as a real grant. The onus is on us to make changes.

Peter Eisenberger: You have correctly analyzed what is going on. There is an underlying problem that you did not mention and that does not get discussed, and that is that universities, too, are behind this. The universities are in partnership with their faculties in emphasizing grant size. They are saying that they want to stress education, but they also look at their books and at the flow of money coming in. Without the money, they can't do anything. So, there is a real problem with coupling education and research, which is the way it is with graduate education. That is why there was some intent in academia to unbundle the financial aspects and the performance aspects. It is a very difficult problem, and you could argue about how parallel these things are.

Soni Oyekan, Marathon Ashland Petroleum: I agree with many of the recommendations from the 1995 MPS workshop, and I would like to suggest they are still valid and may represent some of the recommendations from this workshop. Speaking as someone who has been in industry for some time, I believe that in industry today new graduate students need a multitude of skills. I have a Ph.D. in chemical engineering with specialization in reaction engineering and catalysis. The specific knowledge from those studies was used in the first few years in reaction engineering and catalyst development studies at Exxon and Engelhard. Thereafter, I have had to rely on other skill sets to function in industry. Since the first three years, the skills that have come into play have been those associated with coordination of technology management and support for a variety of oil companies—Exxon, Sunoco, Amoco, and currently Marathon Ashland Petroleum.

So, we can presume that the oil industry is looking for a wide variety of skill sets from Ph.D. holders. These broad sets of skills are in high demand today. In the oil industry, we have essentially eliminated research and development in the main as a result of rationalizations and cost cuttings, and as a result Ph.D. holders entering the oil industry have to come prepared to function in management. Courses such as plant optimization, communication, and advanced process control will be helpful to the incoming graduate students.

David Oxtoby, University of Chicago: When I talk to Ph.D.s from our university working in industry, they say almost uniformly that the most valuable part of their Ph.D. experience was grappling with a real problem. In many cases it was a messy problem of the type you are talking about, one without definite solutions and without a timeline that would enable one to finish in a specific time period. I would say a couple of things about your comments. First, I think the distinctions between research and education are false and that the whole research process and working together in research is education at the graduate level. I think that trying to draw lines and saying, "We are doing too much research and not enough education," is really misleading. Second, I would agree with you that there should be more emphasis on team work and on skills such as writing and communicating and that these should be done more systematically at the graduate level. Third, in terms of time, I think it is unrealistic to fit things into narrow timelines. One of the consequences will be, and is already, more and more expectations of postdoctoral experience, not only in academic positions but also in industry. If we shorten the Ph.D. program, we will have more and longer postdocs. I am not sure that is a step forward.

Peter Eisenberger: I agree with you that a strong separation between graduate education and research is not right. However, to say they are the same is also not right. I think the intent in the workshop was to show that the distortion seems to be the greatest if learning is not educationally bundled but is essentially research bundled. That affects the time. I think it is a difficult question, and it is not black and white. It requires looking at the part of the research programs that is destroying the educational factors and modifying the research activity so that you don't lose it.

2

Graduate Education in the Chemical Sciences

Edel Wasserman
E.I. du Pont de Nemours & Company

With the fluidity of the industrial landscape, as companies merge and restructure, as one relevant science matures to be replaced by another rapidly developing field, it is appropriate to ask how graduate education in chemistry functions in a rapidly changing environment. Here I focus on graduate and postdoctoral education as a prerequisite for a technically based career in industry. Given the many varieties of education and their many roles in the workplace and in professional development, this discussion deals with only a few issues.

- It is important for the scientist newly employed in industry to be able to make a contribution within months of joining the laboratory. Some specialized expertise to be tapped is normally necessary in these early months. However, the nature of industrial needs can change over a period of years and the individual has to adapt to new opportunities. To this end, the graduate curriculum should contain a broad range of science experiences.
- In the graduate and postdoctoral years, the students should be exposed to chemistry broadly through classes, seminars, and opportunities to participate in outside meetings, as well as being encouraged to read broadly outside their own thesis or research topic. Balancing the needs for the student to be productive in research with the longer-term educational needs for the student in a later career is a challenging task for the research director.
- Students must realize that chemistry in many areas is a young science. While we have made great progress in specific fields, closely related domains provide vast opportunities in both the fundamental science and in applications that may be of particular interest to industry. An interdisciplinary thesis can demonstrate how existing knowledge from different fields can be combined in a new intellectual synthesis.
- In trying to provide this balanced environment one should try to design a program that is appropriate for the good and very good students. The truly outstanding students will find their own way and frequently resist the more formal structures that exist within a department. They will create their own world. All the university need do is to provide the environment in which they select and absorb and work in their unique ways.

• We should avoid a prolonged graduate experience. There is anecdotal evidence that programs that go beyond 5 or 6 years may narrow students rather than increase broadening. It is best to keep the graduate experience limited and then let the students learn more in whatever new environments they face after leaving the university.

• Most important, the student must be taught how to learn throughout a career.

• In industry people have the opportunity to seek out colleagues with complementary skills to tackle problems that an individual alone might not be able to solve. Preparing students for the transition to this more diverse world is another way mentors can make a major contribution to student development.

• As a professional, some time should be spent on absorbing new science in other areas. Given the wide range of interests of colleagues in industry, some of the best educational opportunities are informal exchanges with other scientists. Learning how to describe one's own interests and capabilities, and learning how to question others so as to learn theirs, provides the scientist with a toolbox that can be used effectively for many years. Graduate school is the best incubator for such skills.

We often refer to chemistry as *the* or *a* central science. However, in practice it is effectively a decentralized science appearing as a significant component in what passes under the names of many other disciplines. We should attune ourselves to the opportunities for making contributions on the basis of their content rather than whether they are labeled as one form of chemistry or another.

• In school we are faced with educating a variety of quite different individuals. No two students are the same. Finding ways to play to their strengths and advising them on how to compensate for their weaknesses is an important part of the graduate process. Some individuals in the organic chemistry and molecular biology areas do not take easily to quantitative concepts, but they may develop instead a fine feeling for structural and spatial relationships. The faculty advisor may be more aware of the student's capabilities than is the student. Careful guidance of each person can provide the information that allows entry into the outside world with a sense about where one is likely to be most effective.

• Students should realize that a large number of opportunities are available and will be available throughout the decades of their careers. But it will be largely their responsibility to become aware of them. Increasingly, people must manage their own careers, although advice from many others is necessary before they make their final choices.

• Given the rapid pace of scientific change we are often forced to deal with individuals in midcareer (typically in their 40s or early 50s) who find themselves not competitive with new graduates in specialized areas. This is a great tragedy and one we should do all we can to avoid. By planting the seeds in graduate school we can prepare for a continually developing professional able to adapt, synthesize, and contribute for decades. Industry wants to employ its experienced scientists as long as they remain contributors. It is a great loss for all if someone with 20 years of experience has to leave a company because of being less productive than a new hire might be. The chemist who learned how to teach himself or herself in graduate school is likely to remain a long-term contributor.

Many more specific suggestions could be made as to what would constitute good practices in graduate school. A list of some of these came out of a workshop chaired by Professor Ronald Breslow of Columbia in late 1995, and I've attached most of the "Comment" he wrote summarizing that workshop. These suggestions, if incorporated broadly within graduate schools, will do much to improve the capability, contributions, and long-term effectiveness of the professional chemist.

SUPPLEMENTAL INFORMATION

The Education of Ph.D.s in Chemistry
December 11, 1995, *Chemistry & Engineering News*
(From "Comment" authored by R. Breslow; reprinted by permission)

A conference was held at Columbia University to discuss U.S. doctoral education in chemistry. Participants were invited from major graduate programs and from other concerned groups, such as industry. Many participants are heads of departments, others direct graduate students. They represent some of the strongest university chemistry departments, which produce 20% of chemistry Ph.D.s in the U.S.

Conferees from industry described qualities they find most valuable in prospective Ph.D.-chemist employees, while the academic participants described the content of their graduate programs and any planned changes.

There emerged a remarkable agreement on what doctoral education in chemistry should accomplish and how the goals might be achieved:

- Ph.D.s should achieve mastery of a specific area of chemistry such that they can perform as true professionals. Extensive involvement in research, leading to a thesis, plays a critical role.
- At the same time, Ph.D.s should gain educational breadth that covers chemistry and related fields. This should help graduate students realize how knowledge can be applied in other areas, familiarize them with experts and literature outside their own sub-discipline, and extend their thinking about science. It should also serve as the basis for continued intellectual growth.
- To help achieve these goals, a graduate program should normally involve a full year of advanced course work. In most of the departments represented, one-third or so of this course work must be outside the student's general area of research (for example, physical chemists taking courses outside of physical chemistry) to provide educational breadth.
- Advanced courses need not occupy an entire semester or quarter. Several departments are experimenting with modular courses, each module extending over half a term or less. Breadth of education might be particularly well served if early modules deal with the basics, while later modules serve the needs of the experts. Students should also regularly audit courses outside their own fields.
- Students should attend seminars, and not just in their special areas. Departmental colloquia given by outside speakers should be attended by all chemistry graduate students. Faculty should also attend to make it clear that narrowness of interest is not a virtue. Speakers should be told that the audience is general and that some overview is needed.
- Some visiting speakers should be from industry to expose students to nonacademic research. Close connections between university departments and industry can help students plan and obtain industrial careers.
- Scientists need to speak well. The graduate years should offer several opportunities to deliver public seminars, both on personal research and on topics from the research literature. Students need to learn how to describe their research, and that of others, in order to convey its relevance as well as the details of the work. Speaking at a national or regional chemistry meeting can be a particularly maturing experience.
- Scientists need to write well. Regular written research reports, properly critiqued, will lay the groundwork for a good thesis and for good scientific papers.
- Original written research proposals, defended orally, play an important educational role. One proposal might be closely related to thesis research, perhaps elaborating on ways it could be extended. Another proposal, on a topic different from the student's research, should make it clear that the student's own ideas are involved. Research proposals are critical in interviews for academic positions. They also

indicate creativity, something that is helpful in obtaining an industrial position. These assignments help build the habit of reading the chemical literature, a hallmark of professionals.

- "If you want to learn a subject, teach it." Teaching experience is a valuable part of graduate training and an important service. Laboratory supervision is useful, but graduate students learn even more if they can give laboratory lectures or recitation sections. Departments should make sure that graduate students receive adequate training and supervision for these duties.
- At orientation, new graduate students should receive instruction in good scientific practices, including matters related to scientific ethics.
- Time needed for the degree must not extend too far. Graduate education should not require more than five years. Some institutions cut off support after five years. Future employers are concerned when they see lengthy periods of time spent in graduate school.

Some of the points listed above are properly the subject of departmental rules and norms, but much also depends on the careful mentoring of students by faculty. For example:

- A thesis committee for each student, including the thesis director, should be set up early. The committee members should hear and critique the public seminars and research proposals of students, monitor research progress, and participate in any special examinations required. The members should be part of the final dissertation committee. They will be important references in future job searches.
- Thesis directors have special responsibilities. They must follow the progress of research and furnish advice. They must coach students on speaking and writing skills. They should preview and critique the talk that students will give in job interviews; a poor talk can be damaging. Faculty mentors should recognize that a chemistry Ph.D. degree can lead to a variety of worthy and rewarding careers: they should support the decisions of students who choose nontraditional career paths.

Conferees were pleasantly surprised to see how many of these ideas are being followed already or are being instituted in leading universities. The U.S. graduate programs that produce Ph.D. chemists are quite strong. By considering the above recommendations, some programs can become even stronger.

DISCUSSION

Michael Doyle, Research Corporation and the University of Arizona: Dr. Wasserman, I agree with many of the things you have stated except one. That is, the importance of leadership is diminished. To look for the second tier is what you were implying. We are now involved with an enterprise in which bright students are entering graduate schools and looking for opportunities for career advancement. First of all, they are looking for the best opportunities in the academic and industrial world. Where are they going to go? They are going to go to the places that industry and other academic institutions regard as elite experiences. DuPont, for example, does not come to the University of Arizona to recruit. I don't know the list of institutions at which DuPont does recruit, but this selectivity is true of many companies. They come to certain institutions that they believe will allow them to draw the elite into their enterprises.

Second, when students enter a graduate environment and they see that they do not have the opportunity for the best possible education that would lead to the best career opportunities, they are going to look to other ventures and operations. This is true of a growing number of students who are moving from the Ph.D. program to a master's program with the thought that they will have an opportunity to work at DuPont with a master's degree from institution X, because they are not at institution Y, which would provide them with that opportunity.

The third is that, by not working toward that sense of excellence, by not being attentive to the capabilities of the students that we admit, we are allowing students to obtain Ph.D.s who are then going back into the enterprise with poor experiences. This is not necessarily because of institutional problems in training, but may occur because the students were not well matched to their institutions. They are taking up careers for which there is a continual bemoaning that they do not have opportunities. I look, for example, at undergraduate institutions and the faculty being brought into them. These individuals, who have gone through a graduate program, are saying that research was not the career they wanted to have, and they are entering the undergraduate institution because they want to be good educators and do not care to do any more research. Basically what I am saying is that it is excellence that we want. We have to be selective, and we have to provide good opportunities in a variety of ways.

Edel Wasserman: I agree with you. I want to add the point that many of these students do not know the available choices when they are choosing their careers. Once they have chosen a career in one subject and their advisor is providing the best possible advice, suggesting a postdoc with someone who is a world leader in the field, their flexibility is quite small. This is true even if their minds are open to new ways of thinking, new areas of training, and following a different direction that might provide them with skills that nobody else has in exciting areas.

The need to mentor is critical. Certainly, in industry, we find that people need mentoring until they retire. Individuals are constantly changing. A number of authors have pointed out that every decade or so there are major changes in the way our minds work. Our bodies deteriorate, but our minds change and get better.

I think a number of people end up disappointed by the prospect of spending decades in a laboratory, which is essentially all they see until they leave the university. Some of them solve that problem by going into other areas in industry in which there are many possible career choices within the same company.

What I am asking is that we take a hard look at students as early as possible. Sit down with them and tell them about possible graduate thesis topics that may be something they could work on for the next 40 years because these are new areas and exciting things are likely to continue happening. On the other hand, they may become obsolete, in terms of exciting new developments, in 10 to 15 years. The students will have to be prepared for what to do afterward. They will want to build on what they know. Going into a totally new area where previous experience has no significance will not help them. They need to prepare for continuing development and continuing education.

Students should talk with their graduate advisors, other people in the department, and alumni and one way or another find the information that prepares them for a continually changing world.

Isiah Warner, Louisiana State University: You made a statement, and I am not certain whether it was facetious or not, but I totally agree with it. It is that we should concentrate on educating the very good to good students, as opposed to the superbright. I believe that we often tend to ignore the very good to good students. However, if we look at industry, it is those students who are there representing academia, as well as producing for the betterment of this country.

If it were true that the student had to be superbright to be a productive Ph.D. in chemistry, for example, then graduates from the California Institute of Technology, the University of California, Berkeley, Harvard, and the Massachusetts Institute of Technology would populate all of our academic institutions as well as industry. That is not the case. It is obvious that the good to very good students are also capable of outstanding productivity. In fact, schools such as Louisiana State University are concentrating on those students, like the ones at the University of Arizona that Mike Doyle talked about, and are finding that they are producing a very excellent product. I think that more academic institutions should concentrate more on those students, and I want to applaud you for making that statement.

Edel Wasserman: I was not being facetious, and I also did not mean to exclude the very best. What I mean is that it is almost impossible to produce an appropriate course of formal study for someone with a truly original cast of mind. By making a variety of resources available, the university makes it possible for such individuals to gain a deeper education. We want to make sure that those resources are available.

For example, some of the best people never attend class. I had a research director who was proud of the fact that he cut most classes in school. He would boast about it to us, but he said that if someone cut his class, he would not take it kindly. The world is not always fair.

I believe that we can have a great impact by focusing on the formal requirements for the good and very good students.

John Warner, University of Massachusetts, Boston: We discuss graduate education as preparing students to be successful in industry. I would like to make a distinction between being successful in industry and being hired by industry. We have talked about the needs of industry, breadth, and economic realities, but typically the hiring practices are highly focused. So there is a dichotomy, and there needs to be some way of addressing that.

Edel Wasserman: I agree. One time, years ago, when I was talking along similar lines but in a somewhat different context, someone said that the people from industry talk with forked tongues.

The CEO, on one side, talks about broad training for a 40-year career. The person actually responsible for hiring is concerned with finding somebody to do a particular job now. The net result is that we need a combination of capabilities. A broad overview of many areas of science combined with considerable experience within one or more specific areas is a powerful combination.

On the other hand, when we see somebody who is trained in a different area, but whose intelligence and fundamental understanding is first rate, we make them an offer. We will worry about the details later. We do not see many like that. I agree that we are not consistent. But while there is a difficulty, there is also an opportunity.

I have had people call me about a great candidate who has sent her résumé to two different personnel departments in DuPont, but nothing has happened. I get more information and I take it to the human relations department. I obtain letters of recommendation, which are rare things these days, and I make a case. It takes a significant amount of time, but for a good candidate it is worth the effort.

We have had several candidates come in on that basis. If you know somebody at a company where a student would like to interview, contact that individual. Sometimes, of course, you run into a hiring freeze. Then it is best to wait and try again.

J. Michael White, University of Texas at Austin: My question deals with Moore's law, a law in microelectronics that says that basically you are going to have a new, smaller-dimensioned product every 18 months. My colleagues in the microelectronics industry say that the people we send to their organizations should be thinking of reinventing themselves based on roughly that time scale. Does this fit with your model of people being willing to move about among various tasks that they are assigned to?

Edel Wasserman: There is a limitation on how easily people can reinvent themselves. One of the things I am concerned about is whether we are taking advantage of the possible flexibility in people during the graduate period to try to give them a broader view. The normal outcome that I think most of us have had is that if students see a variety of experiences in their graduate years they are more amenable to interdisciplinary programs. This could mean working with several professors who have a joint grant. It could mean working with one faculty member who has a diversity of things going on in a research group. Once you have a long period of graduate school without that broader view it is more difficult to reinvent yourself.

J. Michael White: I think that is an important point for those of us who are in the graduate education business. We must recognize that there are opportunities for at least part of the Ph.D. dissertation to introduce, perhaps systematically, something akin to a mentoring experience with some other faculty member or some other project. It doesn't actually have to be funded. It can be something written down on a piece of paper, such as, "I will learn this skill in the next 6 months."

Edel Wasserman: I want to mention an approach that has been used in a few departments. They call on graduates who have been in industry and have had varied experiences there to come and spend three or five days in the department. They have lunch and dinner with students and faculty, give a few talks in classes, hold seminars, and so on. The graduates bring diverse backgrounds to the academic community.

J. Michael White: I would say that the effort needs to be extended over a long enough period of time to have some significant impact. Otherwise, it will tend to be like this meeting, in that we will go home and forget about it.

Edel Wasserman: The important legacies from my graduate research advisors were basically three sentences. They didn't necessarily strike home then, but within a few years they did. It is amazing that, if you get people at the right time in the right environment, you can make a very strong impression.

We should continue to reach out to students at all stages of their education. For just the reasons I mentioned, some will respond, others will not, and some will have delayed responses. The one thing we should not do is say it is inefficient and hopeless, and therefore we don't do it. That would be a big mistake.

David E. Budil, Northeastern University: I am pretty new to this game, so I guess I have more questions than comments. You mentioned a study, at MIT, I believe, in which students who finish their Ph.D. earlier tended to be broader in terms of their pursuit of chemical research, and the students who were in school longer tended to have a narrower focus. I am wondering to what extent there is institutional control over this. To what extent does it reflect the program rather than the student's ability? Different people have different styles of learning. It is not clear how much you can do to force students to finish in a certain amount of time and force them to be broad. Obviously, that is a desirable thing. I think I hear general consensus on that, but how well can that be controlled?

Edel Wasserman: You are raising a challenging issue. We have set up the academic system so that individual research directors have essentially total control of what happens within their research groups. There are modifiers in some places more than others. But fundamentally, it is an apprentice system, a system that has worked well but does not change easily. However, beneficial changes are occurring.

An individual faculty member may not see that a shorter graduate period could be advantageous and not realize that keeping the student two years longer could make the student narrower with detrimental consequences down the line. But some departments are limiting the total residency or years of support.

Reaching out to faculty members can be extremely difficult. Let me just indicate one recent occurrence. We are trying to run communication workshops in the American Chemical Society to help students who have never given a talk outside their research groups to reach a broader audience. It is amazing what two 15-minute periods can do to their skills. They know everything they have to; all they need is a little practice. Our biggest difficulty is getting the students to participate, because faculty often do not see the need for such a skill. What we can do is to carry the message back home that such skills are needed. We can make improvements.

R. Stephen Berry, University of Chicago: I want to amplify the two presentations we just heard. A physicist at the University of Chicago, John Platt, tried to find any variable or variables of behavior in graduate school that correlated with later success in research, which was measured by number of publications, a variable Platt recognized as an imperfect measure. The basis of his study was the physics department and its graduate students at the University of Chicago. He found only one variable that correlated in any way with later success: there was an inverse correlation between later productivity and the time spent in graduate school.

His most alarming finding was that the data—that is, the students—fell into fairly well defined groups, which corresponded to the research groups in which they worked, i.e., to the faculty members directing their research!

Lynn Melton, University of Texas at Dallas: One of the respondents raised the specter that decreased time in the Ph.D. program would lead to increased time in postdoctoral training. My conversations with industrial hiring managers indicate that there is no genuine need for postdoc training prior to an industrial job.

The D-Chem program, a structured program to have students finish in 5 years, has a 90 percent direct placement rate from campus to industrial career positions, without doing a postdoc. I think that the argument that decreased time in the Ph.D. program would lead to increased time in a postdoctoral position is not sustainable.

Edel Wasserman: We find that the stated needs for new hires vary dramatically. Some recruiters believe that candidates must have a year or two of postdoctoral chemistry before managers or directors will hire them. Others will say that they have no rules but evaluate each candidate individually. Some students are ready to contribute to industry by the time they leave graduate school. On the other hand, in certain areas of the life sciences in which specialized techniques have been important, there is a strong consensus, at least in parts of industry, that a candidate should have had one or two years of postdoctoral experience.

Lynn Melton: I would say that applies to the pharmaceutical industry more broadly.

Edwin A. Chandross, Bell Laboratories, Lucent Technologies: You began by talking about innovation in industry, a topic of great interest at Bell Laboratories, where innovation is defined as taking an invention and making products out of it. Finding the best scientists to do this makes innovation more difficult than just inventing.

I think that to a large extent industry has created a problem for itself. The best people will not come and stay if you don't encourage them to take part in the overall scientific enterprise. They have to go out and talk about what they are doing, to publish, and so forth. There is a monotonic decrease in doing this within the industrial world today.

Second, many students are not going into the usual employment areas. They go to small companies, many of which are start-up companies. Frequently, they are the brightest students, who see the best opportunities there to have an impact in the short term. It is not only the stock-option benefits that drive them. I think it is the opportunity to really get in and be an important part of what goes on, rather than be a member of a huge team. It is a good thing that we should not discourage.

Finally, I want to quote the conversation that I had yesterday with one of the top scientists in the United Kingdom, without identifying him, although I will say that he occupies one of the best-known professorships at one of the best-known universities. He pointed out that he thought the model was changing now and that it seemed that universities are driving much of the innovation, as opposed to companies. There is a lot of truth to that, and it affects the question of how cooperation with industry should affect research. What do students get trained to do? Do you let them talk about it? How do you patent it? are all increasingly difficult questions connected with trying to get students to work on real-world problems.

Edel Wasserman: To some degree, universities have always been the source of innovation. The difficulty and the opportunity are that the university has become more entrepreneurial. Some of the people who might have been good candidates for industrial research are now faculty with their own research groups, and they have started one or more companies on the side. Very good things are coming out of that. It also means that some of the critical reasons for industrial research are being transferred out of industry. You have to ask, what is the job market going to be for students in some segments of industry if this continues on for many years?

James S. Nowick, University of California, Irvine: Your comment about the good to very good students resonated with me. It is easy to get in the habit of sorting students, like eggs, into grade AA, grade A, and so forth, and I think this is a dangerous thing to do.

A student of mine, an undergraduate, was a low B student. When he applied to graduate school, no one accepted him. I believed he had a lot of potential, so I encouraged our program to accept him. He proved to be a bright and successful research scientist and after four years with me had published 10 papers and obtained his Ph.D. When it came time to look for a job, he competed against people who had postdocs and succeeded in winning the position. As a matter of fact, he was hired at DuPont, and I understand that he is doing very well there.

3

Graduate Education in Chemistry: A Personal Perspective on Where It Has Been and Where It Might Go

R. Stephen Berry
University of Chicago

Graduate education in chemistry, and in the physical sciences generally, has changed so gradually that those of us deeply engaged in it for most of our professional careers almost do not perceive the changes. Nevertheless, during the interval from the time senior members of the community, ourselves that is, were students until now, when we approach the last years of our teaching careers, there have been significant changes that we should recognize. Some are clearly for the better; some, essentially neutral in value, are adaptations to changes in the surrounding society; and some appear to pose serious problems.

I will describe my own perceptions and experiences in terms of the evolution of how I and, to some extent, my colleagues have worked with graduate students since we began. I will even interject a bit of recollection of my own graduate-student experience. Then I will turn to the issues of what is good, what is neutral, and what is worrisome.

HOW HAS GRADUATE EDUCATION IN CHEMISTRY CHANGED?

One evolutionary change in graduate education in chemistry that people frequently cite is the increase in research that crosses disciplinary lines. It is fashionable to establish formal interdisciplinary programs; sometimes it is almost mandatory in order to receive funding. As I reflect on what people were doing when I was a graduate student, I am not altogether convinced that there has been much change in the research, or in the perspectives, of people working across disciplinary lines. Konrad Bloch, Frank Westheimer, and Paul Doty were among a large community working during the 1950s and 1960s in the borderland between chemistry and what is now molecular biology. Linus Pauling's work on structures of biopolymers was certainly of this kind. Atmospheric scientists were comfortable doing all sorts of chemistry in the context of the behavior of substances in air. Chemistry of air pollutants was already such a well-developed subject when Sherwood Rowland decided it was time to do some research in the subject that he had to look for exotic species, and settled on fluorine. The subject of atomic and molecular collisions evolved primarily in physics but, with the work of Sheldon Datz and Ellison Taylor and of Philip B. Moon and associates in the mid-1950s, physics and chemistry developed a field that merged the two disciplines. Further from chemistry, in hydrology and geography, Gilbert White was paying no attention to the traditional bounds of any academic subject as he studied how we use and might use water. In other words, I would like to challenge the dogma that science has increasingly become more interdisciplinary. What I propose is that we have merely found it fashionable to institutionalize cross-disciplinary activity that was already happening and evolving in a healthy manner.

Suppose we grant that my challenge contains at least a little truth. What do we accomplish by institutionalizing interdisciplinary research? One change I suspect has happened, but would find it very difficult to support with hard evidence, has to do with another kind of change closely related to this interdisciplinary issue. While people were working across borderlines 40 years ago and more, other people were working in some subfields of chemistry that had become "mature" in the most pejorative sense of that word. In that period, many of my contemporaries and I looked on classical analytical chemistry and electrochemistry as fields supported by an ingrown, myopic community that had lost a larger vision. Analytical chemistry evolved out of that slough and became a much livelier subject when it looked beyond its traditional boundaries and decided that those boundaries had become irrelevant. Rather, the fundamental questions of how to make reliable determinations of what was present and how it behaved became again the focus of the subject.

Another subject that collected specialists who turned from fundamental scientific problems to solving challenging puzzles was, for some years, the synthesis of natural products. It, too, became a "mature" subject with its own rules. When Gobind Khorana used bugs to synthesize intermediates for his eventual synthesis of Coenzyme A, he violated the rules of the game and was considered déclassé for doing so. Again, synthetic organic chemistry has outgrown that stage and is again a healthy science, in part because its practitioners have looked outside traditional disciplinary bounds and flirt regularly with materials scientists. In what field do they work? I consider that an irrelevant question. It is more important to ask, What interesting and important problems are they addressing, and have they the skills (or access to the skills) appropriate to those problems?

One other "mature" area comes to mind that, according to my prejudices, has not yet emerged from its slough. This opinion will raise hackles, notably among its practitioners, who have been slow to recognize the maturation. This is the area of quantum chemistry, the application of computational methods to determine the properties of atoms and molecules (and sometimes, for the more interdisciplinary sorts, of solids and even liquids and polymers) by solving quantum mechanical equations. I would certainly not deny that people in this field have been making progress with the methods for doing such calculations. What I would raise is the question of whether the problems being addressed by that rather closed community have evolved into problems "within the club" and have become problems whose answers are of little interest to the rest of the scientific world. Can that community explain persuasively why their achievements are interesting to people working in materials science, or molecular biology, or atmospheric chemistry, or fundamental quantum mechanics? But there are fundamental, important, and potentially productive questions standing open in this field. Can anyone recast the formally exact density functional theory developed by Pierre Hohenberg and Walter Kohn, starting with first principles, in a form whose lowest approximation is the ubiquitous (but still only ad hoc) local-density approximation, that would also reveal how to make further refinements of this powerful approach?

Thus, I infer that interdisciplinarity has had a healthy effect in pulling "mature" disciplines back into true scientific import. Institutionalizing the interdisciplinarity is a way of making it financially attractive to look beyond traditional bounds. Whether such institutionalizing has also increased true intellectual interactions is a question I cannot answer, but as I indicated, I am somewhat skeptical that it has. To establish whether it has would require a serious survey, and even if conducted the survey would be suspect on the basis that it might be trying to find a particular answer. Certainly we can find many examples of productive institutional interdisciplinary structures, and we cannot expect to find people eager to point out cases in which having the institutional structure has not been successful. Unfortunately, those less successful examples, the cases people are reluctant to discuss, are the cases from which we would learn the most.

If we examine how institutional structures affect the scientific enterprise, we must be prepared to go about it objectively, i.e., we must look for both the positive and negative consequences for science and education of institutionalized supporting structures with specific goals. We must look at not only those structures targeted at interdisciplinary work but also those designed for traditional disciplines. Are we better off with designated, targeted institutions, or with an informal, fluid enterprise? Do narrowly defined structures such as "centers," whether funded or unfunded, contribute more or less to science and science education than larger, more broadly oriented, fluid organizations?

Two more of the most apparent (and related) changes in the U.S. chemistry graduate programs have been the increase in the length of time from entry to doctorate and the increase in the length of doctoral theses. Typical graduate careers, B.A. to Ph.D., lasted about 4 to 4 1/2 years in the 1950s and 1960s; now 5 to 6 years is more typical. The doctoral theses on my shelves were of the order of 80 to 130 pages into the 1970s but are typically about 200 pages or more now.

These are symptoms of change. What changes have been responsible for them? An obvious and rather trivial contributor is the use of the computer as a word processor, so that the cost of doing a long thesis is negligibly greater than the cost of doing a brief thesis. This contrasts with the situation of having to pay a typist, on a per-page basis, to put a thesis into acceptable form. Furthermore, it is vastly easier to write and revise a text with a word processor, so that we might expect the writing quality of the theses now to be higher than it was in the 1950s and 1960s. I am not convinced that has happened, but I hope someone does a doctoral dissertation on that subject so I can find out. One thing we can expect to improve with computerization of the scientific literature, something just now at its threshold, is the quality of the scholarship, specifically the accuracy of attributions and historical background. As the archives of journals are converted into searchable electronic forms, we can justifiably expect students to find and cite early works directly relevant to their research. The barriers to doing such searches have been significant, so much so that we have all heard outraged criticisms of how the literature has been overlooked, especially literature in languages other than English.

Another reason often given for the growing duration of the graduate career is the need to know much more than was needed 30 or 40 years ago. Certainly the body of scientific knowledge is vastly larger than it was in 1950 or 1960. We sometimes rationalize 6 years of graduate work on the basis that a student finishing a Ph.D. must know about the same fraction of the literature that was expected of a fresh Ph.D. in 1960. Hence, we argue that students must spend more time in school to absorb all that material. If this is so, should it be adequate justification for the increase in time to the Ph.D.? This takes us past the realm of identifying changes into the subject of the next section dealing with what graduate education should achieve.

The collaborations between universities and private firms have been growing rapidly and often involve graduate students. This has been a significant change in graduate education in chemistry and in the relationships among institutions and individuals. The history of this involvement is tied closely to the evolution of the research industries in America and to the perceptions in Congress of the role of government in assisting the evolution of science and technology. As a result of the apparently transient fad of industries divesting themselves of much of their longer-term research programs and seeking outside sources to do that research, the universities appeared to offer natural, low-cost alternatives. The most obvious symptom of the mood of Congress was the passage of the Bayh-Dole Act, which allows patenting of results of research done with government grants. The result has been a willingness of university scientists and university administrators to establish collaborations with industries. This has led to a number of changes that are discussed in the final section of this paper.

WHAT SHOULD GRADUATE EDUCATION ACHIEVE?

Let us return to the questions stimulated by our asking why graduate education programs in chemistry and other sciences take significantly longer now than 40 years ago. We may begin with a traditional view of the course of a career in science. The graduate student is probably as close to being a traditional apprentice as anyone in modern life. The graduate student works under the direction of a master, who guides and trains and, we hope, educates the apprentice until the student crosses the threshold of an advanced degree, normally the Ph.D. The student-apprentice has become a journeyman.

Here we can clarify a common misunderstanding, the bifurcation that is sometimes made between education and research. In truth, all true research is education in that it is neither simple training, if it is under the direction of someone else, nor a pursuit divorced from any impact on the researcher. Education includes any kind of experience that leads us to modify our behavior. It includes more than research, e.g., pedagogical classroom processes, but certainly the process of conducting research causes us to modify our behavior and our thinking continuously, as we carry out our investigations.

The journeyman stage of a scientific career may be unclearly defined if the person goes immediately to an industrial or government job, but even there the newly hired scientist is likely to go through a sort of training period or interval under some supervision. If the fresh Ph.D. goes instead into academia, then the journeyman stage is easy to recognize as the period between receipt of the doctorate and the achievement of tenure. One change that has happened to this part of the career path between 1950 and 2000 is the transformation of the postdoctoral stage from being exceptional to being almost mandatory. That transformation occurred about the time of the great expansion of support for science, the period following Sputnik. The journeyman stage now happens in two stages—that of the postdoctoral, even of two or more postdoctoral periods in different groups, and that of the junior faculty member. The journeyman is expected, as in medieval society, to be able to produce work that may well be of the same quality as the master's. But the journeyman is under a kind of scrutiny and evaluation that the master, the tenured faculty member, does not have to endure.

Graduate education, then, turns a novice into an apprentice and eventually into a journeyman scientist. At that stage, we expect the fresh Ph.D. to be in full professional command of at least one subject, perhaps a very narrow one. More precisely put, we expect that person to know more than anyone else in the world about the specific subject of the doctoral dissertation. We also expect that the experience of doing the research for the dissertation, and then writing and defending the dissertation, have educated the person to the stage of being able to invent or recognize, and then pursue, new research problems to their conclusion. This implies that the Ph.D. education included learning how to teach oneself new subjects and to do this well enough to use those subjects in new research.

We also expect those graduates who continue in academia to have learned something about teaching science as part of their apprenticeship. Most but not all graduate students now spend some time as teaching assistants. I think every scientist would agree that there is no better way to deepen one's understanding of a subject than to teach it. The teaching experience is not only important for those who become professors; it is also a part of the apprenticeship that strengthens the foundations of insight for all graduate students.

Graduate education serves another purpose that I would like to believe fits neatly with the apprenticeship. That purpose is to work with the research director on problems that are part of that faculty member's research program, with the goal of furthering that research. We in the world of academic natural science believe that working on problems that are part of our research programs is an excellent way for graduate students to learn the process of doing research, as well as for helping us investigate problems we think are interesting and important. (Incidentally, neither mathematics nor the humanities operate this way. It is far more usual in those fields for graduate students to choose problems of their own, which may be unrelated to any specific interests of the faculty supervisor.)

This apparent symbiosis deserves critical examination. It may be that it is indeed a healthy and productive educational pattern. It is also possible, however, that it has some weaknesses and flaws. Pressure bears heavily on faculty members to raise the grant support required to carry out a research program. "Produce!" is the cry, and producing results is a necessary if not sufficient condition to obtain funding. This means that faculty members may be subtly tempted to ask for (or demand) a considerable body of research results from students before allowing them to submit their theses. This temptation, if it exists, is obviously greatest when the student has become skilled in the research. We must look hard and critically at our behavior and ask whether some of the increase in the time spent in graduate school could be caused by demanding more of our students than our mentors expected of us. That in itself, though, is not necessarily good or bad. If we, however, answer "yes" to that question, we must go to the next question, "Are students aided or impaired in their careers, their productivity, and their creativity by our having those greater expectations?" I raise these issues not because I know the answers but because we, as a community, are obliged to find answers if we are serious about finding better ways to educate our graduate students.

WHAT DO UNIVERSITIES PROVIDE TO INDUSTRIES?

We know the answers to the reverse question. Industries provide employment in which graduates can find satisfying careers, and they sometimes provide financial support for students or for research in graduate schools. These two, and the way I have phrased them, lead us directly to an important point that we must keep clearly in mind. *The primary product that universities provide to industry is skilled, educated people.* The young scientists who take their Ph.D.s to their new industrial posts are superexperts in narrow fields that may have nothing directly to do with their new responsibilities. That narrow expertise, however, is almost never what matters to the new employer. What matters is the ability of the young scientist to meet new problems, to learn new material, and to have the judgment required to decide what to keep of old methods and what to introduce that is new. These are the qualities that make our graduates valuable to the firms that employ them, and they are the qualities that make those graduates our most important product.

Yes, sometimes a secondarily valuable product is the result of the research done in the university. This has become more valuable with the Bayh-Dole Act and the opportunity to capture private benefit from the results of public funding. This attraction, and the drift of academic interests in some areas toward problems whose answers may take the form of, or lead directly to, profit-making products, have led to a significant increase in collaborations between industries and academic researchers. In some subjects, these collaborations involve consortia of private firms working with one or several universities. In other subjects, the collaborations are generally between a single firm and a single research group. The former tend to operate in about the same manner as academic groups that have no such collaborations, following the same sociology and rules of behavior as university researchers have for many years. The latter tend toward industry-like behavior, meaning that proprietary concerns play a larger role in the one-on-one collaborations than they do in relations involving consortia.

The inducements are strong on both sides to establish collaborations. From the industrial perspective, the universities provide fresh talent from the students and high skills (and even sometimes high ability) from the experienced faculty. They also provide cheap labor. Students are paid far less than industrial researchers. The collaboration with a strong university group is a low-cost way for industry to outsource research. On the other side, university researchers and administrators see industrial collaboration as an attractive way to get funding for research. Hence, both parties are eager to collaborate.

What consequences may such collaborations have beyond the immediate ones seen by the participants? In the case of the open-style collaborations with consortia, probably only the beneficial consequences of educating the students to the industrial perspective and, if the industrial side also contributes participants, some education in the other direction. On the other hand, those one-on-one collaborations that operate under the restrictive rules of proprietary secrecy bring into the university something foreign to that environment. Secrecy and refusal to communicate and discuss research work are inimical to the functioning of a university. The stories of research groups in the same department whose students cannot discuss their work with one another illustrate the way the rules of proprietorship can erode the very heart of the academic experience. Open, candid discourse is essential for the education of the students, who are our primary product. We must find a way to establish and accept a set of rules of self-governance that will keep alive the productive industry-university collaborative interactions and in a way that maintains harmony with how universities must function in order to attain their primary purpose. We must not lose sight of our greater responsibility in the zeal to achieve second-level aspirations. Yes, each of us as a scientist considers our own research the most important thing we do, but we act not only as scientists. We are also professors, meaning that we are teachers, with a primary responsibility to educate. Only in an atmosphere of candid scrutiny and criticism, of open discourse, can we maximize the fulfillment of that responsibility. That is how we shall achieve the goals of education that Peter Eisenberger, Edel Wasserman, and, in his comments, David Oxtoby have set for us.

DISCUSSION

Ronald Breslow, Columbia University: All three speakers in this session have referred to the issue of time limits for obtaining a Ph.D. I think this group has got to search within its hearts to find out how we ended up with an unlimited Ph.D. program. We say science is moving, so there is more to learn, and therefore we need more time. The field of medicine is also moving; however, it still takes only 4 years to get an M.D. Whom are we kidding? It should be possible to complete a Ph.D. degree within a well-defined timetable. At Columbia, we have a limit of 5 years on our Ph.D. degrees and that is it. That was done, in part, to protect the students and, in part, to get the students moving. A lot of students feel comfortable in graduate school. They need a push to leave. Now, they know they have got to be out in a certain time. They had better start thinking about their future, planning their postdocs, or starting to interview for jobs.

Frankly, the time limit is also there to protect the students against some faculty—faculty who are no longer with us, who kept students on for 7, 8, or 9 years. It was preposterous. It is not fair to the students. The idea that you establish a well-defined time makes sense in every respect except one, and that is the faculty's own self-interest. We have to worry about that motivation. John Kenneth Galbraith, a well-known Democrat, once defined modern Republicanism as an attempt to give a philosophical justification to selfishness. Let's make sure that we do not do that here, by arguing that there is a philosophical reason for the Ph.D. program without a time limit.

Second, we talk about how to stimulate innovation and creative thinking. There are ways to do this, and we have to make sure that we expose the students to them. Most institutions that I know, but not all, require original research proposals from the students. Those original proposals have two functions.

First, they give students a sense that there is an area that they know more about than the faculty, because they have looked into it and have come up with ideas. I had one student who went to a major university with a research proposal that she had made. She pursued that research proposal in her independent research career, got tenure, and is now one of the major figures in an Ivy League university as a result of that proposal. This is not an unusual situation. The requirement of a research proposal also stimulates students to deal with science in a whole new way. They attend seminars, not passively listening, but thinking, what can I do with what I am hearing? It stimulates them to listen; it stimulates them to read. It is an incredible educational device, and it is preposterous that everybody doesn't use it.

We also require that a student give a seminar on a topic not related to their research to the whole group, faculty and students, within, say, the organic division. They have to look into the literature to prepare for this, and that is often the background of the research proposal. Seminars that they give are not just practice in speaking. They also give the students a chance to extend their education and become an expert in something.

One of the speakers from industry said that what he had learned in graduate school turned out to be useful for about the first 3 years, and then he went on and he needed other skills. My Euclidian geometry probably didn't even last that long, in terms of my application of it. As a component of education, however, it was absolutely critical. I think what we learn in graduate school is absolutely critical as an education: you can take a problem that looks impossible, address it, and solve it. That experience is important. I would not discount the role of graduate education, even if, in industry, people go into different areas. They can go into them with full confidence that they are competent scientists, as good as anybody else in the company, and are able to relate to anyone on an equal basis because they have a solid education behind them.

Finally, I want to raise the question of fellowships. Fellowships are important, and they have been cut back significantly. Steve was talking about what life was like when we were in graduate school. Many people had fellowships. The result is that you had your own money, so you could pick your own sponsor without asking whether the sponsor could support you. A fellowship means other things, too. Fellowships are a signal to the students that they are winners, that they have been selected as having the abilities that promise future success. I don't think you should discount that. We worry about why more Americans don't go into science. If they were winning fellowships to go to graduate school, it would be a tremendous stimulus to them, compared to going in and asking a faculty member to support them. It is a different kind of mind-set.

P. Wyn Jennings, National Science Foundation: I wanted to add a few things, particularly since I am managing the Integrative Graduate Education and Research Traineeship (IGERT) program. One issue that I would like to bring forward is globalization. One of the major concerns is the lack of globalization in American graduate students and whether or not U.S. students are competitive internationally. We know that they are competitive, at least in my field of chemistry.

Another point is diversity. I want to emphasize that this is a major issue from several points of view, both practical and philosophical. If you disenfranchise any segment of the population, it has serious and national consequences. One final point I want to emphasize is that IGERT is an experiment at the National Science Foundation. We are asking you to experiment with graduate education. We have no prescriptive rules other than to be multidisciplinary. We want you to experiment.

R. Stephen Berry: Let me respond to two particular points that you raised. First, from my perspective globalization is not a serious problem, because I see a vastly more international population now not only at the postdoctoral level but also at the graduate-student level in the United States, and it is not just postdocs who are going to stay in this country but also those who are going to return. Our graduate students now are interacting closely with many people from other countries. I don't think we have to worry too heavily about that.

On the diversity issue, I don't think this meeting is an effective forum in which to field that issue. Graduate school is too far along on the education path for us to ask about reasons for women and minorities choosing to study science and engineering. The real choice occurs when they go to college, or earlier. What is it that turns people off at much younger ages? At the graduate level, we are already dealing with a small pool that is much smaller than most of us believe that it ought to be. While we can talk about it here, whatever we can contribute would be a Bandaid compared with the real problem.

Joseph Francisco, Purdue University: In all issues that we have been discussing about graduate education, we have assumed that graduate students are coming in with the right kind of basic knowledge and skills to start with. We also assume that the bottleneck in the whole process is research. This might be true for a few schools—such as Columbia, the Massachusetts Institute of Technology, the University of California, Berkeley, Harvard, and the University of Chicago. However, for a number of other schools, extending the time line is caused, in part, by students filling deficiencies in basic chemical knowledge. Clearly, it is not just an issue of research. It is an issue of getting those students up to speed by developing a basic background before they can do research. So, as we shorten the time line for the Ph.D., are we in fact allowing these deficiencies to go unaddressed? We are creating a pipeline. The problems that are not being addressed in high school and not being addressed at the undergraduate level are soon not going to be addressed at the graduate level. In the long term, I think we will diminish the quality of our graduates.

I would like to express a worry that comes from personal experience. I had a graduate student ask a question of freshmen at Purdue University who are taking general chemistry. (At Purdue, the requirement for taking general chemistry is a year of chemistry in high school.) The graduate student poured water from a tap into one glass and water from a bottle of Perrier into another glass. She asked the students which one was a chemical. A significant number could not answer the question. This is what is being propagated in the system. I think we have to be careful.

R. Stephen Berry: Dr. Francisco, I think your point is well taken. At one time I remember saying—I think it was in my first real teaching job—that there seemed to be an inverse correlation between the number of courses required of graduate students and the quality of the graduate school. The reason is exactly what you just pointed out. It is a question of how much help graduate students need before they can start the research level of education. I think that is one of the reasons why I believe we should not make a change immediately. Let's look very carefully at what function those extra years are serving before we decide to change.

John T. Yates, Jr., University of Pittsburgh: I was intrigued by the comment that chemists and physicists work on problems that are invented by their mentors, whereas people in other fields work on problems that they themselves invent. I would maintain that the criterion for the end of the Ph.D. should be that students are able to invent their own problems or subproblems within the field of the research. Unfortunately, it takes different students different lengths of time in different types of problems to achieve that level of competency.

R. Stephen Berry: Certainly, in a field like experimental chemistry or physics, it takes not only a knowledge of facts but also a certain amount of judgment about how difficult the actual experimental work is going to be and how long it will take. There are solid reasons for those differences.

Bettina Woodford, University of Washington: I am with the Pew Charitable Trust project "Re-envisioning the Ph.D." I seem to be an imposter here, because I am a linguist and not a chemist. Our project is cross-disciplinary, so I have come to this conference. I want to ask a question about how faculty buy into all the change that is being proposed. We heard earlier today about the importance of off-campus practicums for students to get them more involved in softer skills, communication, multidisciplinary teams, awarding funding directly to students rather than to the professors, and shortening the time to degree. These are all issues that our project has heard in more than 300 interviews that we have done around the country, across disciplines and sectors. Another thing that we have heard is about getting students off campus into practicums with industry and getting more faculty involved in off-campus research partnerships. This allows the faculty to be more aware of what students can count on as career options during their graduate training. This certainly varies by discipline, and I am aware that in the sciences it is more of an option than in a lot of other disciplines, such as mine, for example.

We also heard today a concern about how giving funding to students can affect faculty negatively, especially with regard to tenure. Ed Wasserman says that he finds faculty often resist sending their graduate students to some of the softer skills courses and seminars. One of the major concerns expressed by the people we have interviewed is that, for major changes in restructuring graduate education to really work and move forward, the faculty need to buy into them. As we head into the 21st century, do you have any suggestions about what key motivators might get more faculty to buy into this process.

R. Stephen Berry: That is a wonderful charge. Your way of putting it actually fits very well with many of my own prejudices. When I have had students decide—after graduate school or after a postdoc—that they want to go to law school, I think that is great. At this point, our society needs people who know and understand science to work in areas of legislation much more than it needs scientists. I don't have a good answer for how to persuade those of my colleagues who would like to generate clones that science is not the only career path that could be productive or valuable for students. I am not an institutional guy. You probably got that sense. I much prefer having things fluid and flexible and adaptable, with a minimum of rules. I would like to offer students opportunities rather than requirements. By the same token, we should also ask, Why shouldn't industries be flexible enough to allow industrial scientists to take sabbaticals and work in universities or, for example, in government labs? I had this discussion, for example, with some of the Department of Energy labs, places that would like to have more exchange. The problem is that group leaders tend to be resistant because they don't want to lose the people in the group for even 6 months or a year. But that flexibility would be very desirable.

Billy Joe Evans, University of Michigan: I would like to mention that Steve Berry was on the faculty at Chicago while I was there. Even though I never took a class from him, he had a significant impact on me. He was always a flurry of activity in his lab in Kent Hall. For me, that was one of my models of an academic scientist. I am appalled by our powerlessness to deal with the problem that has been expressed here, by leaders in industry as well as by leaders in academics. I agree fully with Professor Berry's notion on keeping things fluid. We have talked about limits, and we have talked about capabilities. Those are not really the things that we are trying to get at, and therefore I would like to move in another direction.

The only reason that research should be conducted at a university as part of a graduate education program is for the sense of power that a student gets in a field that comes from research activity and that can only come from that activity. Until the student generates his or her own equation, or understands a relationship that has not yet been explored or understood, that student has no sense of his or her own power to move a field. So, one is not concerned about capabilities such as whether I can talk or whether I can read or write. If I have a sense of power, I will do whatever it takes for me to move this field. What that suggests is that the kind of research that we do in a university setting is different in character from the kind of research that is done in industry. Our research must be chosen so that when a student has completed it, there is a sense of power that that student has acquired relative to almost any inquiry, but certainly an inquiry of that field.

Research groups that have 20 or more members pose serious difficulties to the faculty's ability to be sure that each student has a sense of power about this discipline. I do not think that such a thing is impossible, and I do not believe that it has to be done everywhere by everybody. I think that the goal of graduate education in the chemical sciences has to be to give students a transcendent sense of power when it comes to dealing with matters of a chemical nature.

R. Stephen Berry: I would agree very, very heartily, and I would phrase my own perspective of that slightly differently. A goal of graduate education in science ought to be that each student reaches a stage of realization that he or she has contributed something to the world's knowledge that is permanent. I think that is perhaps the source of that sense of power, that you have done something that, unless all the libraries are lost, has added to the world's knowledge. Whatever epsilon it may be, you have done it.

Panel Discussion

Peter Eisenberger: I want to establish a focus, since I have been through this process before. The questions that occupied the previous discussion were: Whom are you educating and for what? and Whom do you expect to pay for that education? We can have more than one answer, but they at least have to be self-consistent. That is, if you see what the students you are educating are doing afterward, you should be preparing them for it. If, on the other hand, you prepare them for what you theoretically think they should do, you often end up with a high mortality rate, as they don't necessarily do what you spent money training them to do. Also, it is important to recognize that you can have your own idea of what education ought to be about, but you should also be sure that the sources of the money for it—the institution and the taxpayers—are, in fact, going to agree. If you can openly say, I am trying out a student—I am spending money on this—they have to agree. Unfortunately, if you spend $250,000 or $1 million on a student—I don't know what the dollar amount is for an experimentalist—who then leaves to be a stockbroker, that is a hard thing to sell to taxpayers.

R. Stephen Berry: It may be good for society.

Peter Eisenberger: It is a hard thing to sell to the taxpayers. We ought to be more efficient than that.

Isiah Warner, Louisiana State University: I want to make three quick comments. First, I believe that we have to be careful about the analogy we use for where we want to go. I don't agree with Ron Breslow that medical school is the appropriate analogy. I am old enough to recall when doctors would say to a patient, "I am God and I know everything." Now they say, "I am one of many deities. I am the god of endocrinology," or "I am the god of podiatry." Medical school has become much more specialized.

I totally agree with the comments made by Joe Francisco. There are students that we have to be prepared to train. The best student I ever had is in my group right now. She came in, passed all of her qualifiers, and is doing all the things necessary for her to complete her Ph.D. in three years. At the same time, I have had students who have stayed in my group for as long as six years. I think we are doing a disservice if we set a specific timetable and say that all students have to graduate during that time. This is not going to work, particularly for the students who are not in Chicago, Harvard, or Caltech graduate programs.

The other thing is that we keep clamoring for a model in terms of science education at the undergraduate level. We don't have to go far from here to find a model, at the University of Maryland, Baltimore County (UMBC). They have undergraduates who come into science and, 4 years later, 50 percent of these students are still in science. They have documented data regarding comparable students who came to their institution, interviewed, and elected to go elsewhere for various reasons. As I recall, only 3 percent of the students who went elsewhere are still in science. Clearly, this is a model that works, so I don't know why we continue to say that we have yet to find a model. The UMBC model works. Go over there and let Freeman Hrabowsky show you why it is working.

Edel Wasserman: I would like to come back to the point that Peter and Steve raised. We are dealing with individuals, and they are not going to fit into a single mold. The number of people, however, who will study molecular beams and then become stockbrokers is a sufficiently small fraction that we should not define public policy on such instances.

Most of the students we are discussing are hoping to have a career that is science related when they leave graduate school. They may change their minds. One of the reasons for a change is that through the graduate period they usually are exposed to only a few of the opportunities for those with a science-based education. If they saw a broader view, some might decide, for example, to stop with a master's degree in chemistry and obtain an M.B.A. to allow them to pursue other possibilities. The point is to be flexible. Individuals vary, and one answer does not fit all.

The criterion that someone ought to be able to choose an appropriate research problem as a requirement for a Ph.D. is, to me, unfortunate. Some of the most talented, valuable, and excited individuals I have seen in industry are incredibly good at solving problems but seem to have less interest in choosing broad research areas. Again, I believe we ought to keep a variety of possibilities in mind as we ask what education should be.

James D. Martin, North Carolina State University: Before I make my comment, I would just like to remind us that women are getting graduate degrees, and we need to include "she" in our vocabulary. The issue I would like to raise focuses on the relation of junior faculty to the discussion at hand. I think if you take a look at your junior faculty, you are going to see that many of these issues such as time to degree and interdisciplinarity are already being addressed. I don't know of many junior faculty who keep students in their labs more than four or five years for a Ph.D. Maybe it is because we are trying to get tenure, and we have a clock. The clock on us affects the time for students. Maybe we need to put a clock on senior faculty.

In terms of interdisciplinarity, most junior faculty that I know are interdisciplinary in the way that they look at science. Stephen Berry noted that in the days of scientists such as Linus Pauling there was a greater degree of interdisciplinary thought than is witnessed today. I believe the pendulum may be on its return. Those of us approaching the "messy problems" today have to be interdisciplinary in order to make progress in our research. At the same time, I resonate with the problem that comes when we institutionalize interdisciplinarity. This institutionalization generally comes at the direct expense of junior faculty. It takes a while until junior faculty are known in the community and sought out for collaborations. So, unless one is at an institution with an already established interdisciplinary center, junior faculty have little to no access to these resources. The majority of such resources seem to be going either to established faculty or established departments.

Furthermore, this issue of resources has a direct and major impact on our ability to offer a quality graduate education. This morning we heard that, overall, the funding situation is quite strong, and Dr. Wasserman made a comment to the fact that DuPont had money but was short of personnel to do new science. Yet I do not believe the sentiment in the "trenches" is that funding is good. We learned of the IGERT program in which a total of 58 awards was made over the last 3 years (out of 1,344 applicants). While this is a great program, numbers of that size are needed in the chemical sciences alone.

In terms of the availability of resources, I am reminded of a conversation I had with my graduate advisor discussing my pursuit of an academic career. In the mid-1980s, he indicated that one could reasonably expect approximately $100,000 per year in a National Science Foundation (NSF) grant. Today, that amount is the same, yet I now pay my graduate students more than 150 percent of what I was paid as a research assistant, and the NSF postdoctoral program in the chemical sciences has been eliminated. Graduate students clearly see a difficulty in obtaining resources and witness the frustration of their advisors trying to garner funding at the expense of time and energy that otherwise could be used to pursue education and science. I have unfortunately seen several leave science because, if one is to pursue a career with such a major responsibility to "chase money," one can find greener pastures elsewhere.

Some major institutional issues need to change if we are going to create a model of true interdisciplinary science that also fosters education. We must also realize that the strength and uniqueness of America's science and its university system have their origins in the American frontier spirit that built this country, a place where the little person, with an independent will, motivation, and creativity can do great things. That frontier spirit, not the creation and preservation of institutions, is what made the American university great and is what will take us into the 21st century.

R. Stephen Berry: Funding in the physical sciences has not gone up. It has basically dropped, especially unrestricted core funding. It has dropped more in physics than in chemistry. Overall, our support for basic research in chemistry in universities is significantly lower, especially in buying power, than it was 10 or 15 years ago.

Frankie K. Wood-Black, Phillips Petroleum: This morning, I have heard several different talks, and I have heard a lot of the comments. The theme that strikes me when I am hearing these things is diversity. It is diversity in terms of thought, of culture, and of application. We have industry folks who want to break the barrier to work with the universities, and we have university folks who want to break the barrier and work with those in industry. There are different models at different places, and that is what makes our academic system so great. We need to strive to create that diversity and to keep moving in the direction of diversity, because that is what is going to make us take the next leap forward.

It sounds as though, from the talks we heard this morning, there is a problem, that we are in a place where no one is envisioning what it is going to look like on the other side. Those models are different, and we are in an age where knowledge is the driving force. We are going to have to work on diversity, and all these ideas have to be brought together. One of the problems that I am seeing, both in terms of my professional career with Phillips Petroleum and in terms of working with the American Chemical Society (ACS) (I am the current chair of the women chemists' committee for the ACS), is that we are not breaking that barrier and recognizing diversity in all its aspects—culture, thought, and so on. What we are also seeing is that, because we are not recognizing that diversity, we are seeing what I call self-selection. The pipeline is there. We are seeing a lot more people come in. That is apparent from looking at the diversity in this room. You have industry folks, government folks, minority folks, and a number of women in here. This is a very diverse set.

I predict that within the next 5 years we will see that a third of the folks will self-select themselves out of this profession, or out of industry, or out of academia. That is the crux of our problem, because we are seeing self-selection at the graduate level and at the undergraduate level. So, self-selection is another issue that we have to address.

Marcetta Darensbourg, Texas A&M University: I would ask, is there one thing that we could do to make a real difference? A few years ago NSF asked that new faculty members include a statement on education in their young-investigator grant applications. Those statements have evolved over the years—eventually into sort of a teaching philosophy—and are required in other grant applications now. In addition, those statements eventually turned into part of the college and university application package. The applicant must now include a statement about teaching in a research proposal. This has been good for teaching at the undergraduate level. It sent the message that teaching is important. Everybody knows the rules of the game, but the message says that teaching is important and your attitude toward this responsibility is important.

I think all the mechanisms needed to educate a student broadly are already present in graduate school. We have student seminars and presentations, research proposals, and so forth. The question is, Are our faculty really serious about mentoring the students in or through those mechanisms? On the average, current young faculty seem to be more dedicated to this than are the older ones. Maybe that is the main thing—all faculty should be encouraged to express their philosophy of what the education process should be for our graduate students. We need to express it as a reminder of our responsibility.

Ernest L. Eliel, University of North Carolina: I have two short comments, but before I start, I would like to agree with Dr. Darensbourg and Dr. Martin, my colleague down the road. I find our young faculty to be more flexible and receptive to ideas than the older ones. First, on the matter of globalization, I agree only half with Steve Berry. It is a one-way street. The Europeans come here, and they return globalized. But we Americans live on a big island, and very few students go abroad except perhaps later in their careers.

A good program at the undergraduate level is run by Professor James Boggs at the University of Texas at Austin, which allows undergraduate students in their junior or senior year to go to Europe. It is an excellent program and well organized. The students are slotted into appropriate science courses abroad, and their grades are accepted at their home institutions so that they don't lose a year. Unfortunately, participation in the program among universities is minimal. I think that is a great pity; the program could be used to a greater extent. I think the undergraduate years are the best time to get globalized, when you are young and receptive.

The other point I want to make is on the matter of time to complete a Ph.D. If there is a rule that after five years you lose your financial support, it means that many graduate students will work Saturdays and Sundays. It also means that both the professor and the student know that there is a five-year limit and they had better do something about it—that can have a beneficial effect. We make exceptions when they are justified. If someone comes with insufficient preparation, we might allow an extra year; we usually decide that at the beginning. Also, if the research takes a little longer than anticipated, we will allow one more semester, but not more than one. It can be done; it works for us.

R. Stephen Berry: In response to your challenge about globalization, certainly it is more one-way on one side. I think there are many opportunities for American students to go in the other direction. My teaching assistant this quarter walked into my office before I left Chicago to tell me that he is going to Hamburg today to do some experiments. One of my graduate students is currently spending a couple of months in Madrid. But American academics are probably not as ready as Europeans to use these opportunities.

Edel Wasserman: As to this question of best practices, I want to refer you to *Chemical & Engineering News*, which Ron Breslow produced in late 1995, after a workshop that he ran early in his ACS presidency. It was a summary of best practices as agreed to by a group of department chairs and a couple of people from industry.

James S. Nowick, University of California, Irvine: We all agree that there are tremendous changes happening, and we need to do something about graduate education. Knowledge is growing exponentially. We are looking at this fact and asking how to deal with it. We must recognize, however, that many social and political changes are sinusoidal, rather than exponential. When we look at addressing graduate education, we need to make sure that we are not trying to apply exponential models to sinusoidal issues.

Victor Vandell, Louisiana State University: I am a graduate student, and I might be the only one here. One of the problems I foresee with trying to fit everyone into a single formula to solve the issues arising with graduate education is the individuality of people. Specifically, different personalities and diverse backgrounds will make any solution difficult to impose. I would like to see a more external approach taken to address the issues of how we solve some of the problems in graduate education and generate more interest in graduate education in the next century. That approach may be through marketing.

Why is it not an option to improve the marketing of graduate education in the chemical sciences? This would excite young people about the discipline and encourage them to pursue advanced degrees in these areas. I don't see that happening. I think this area of the chemical sciences rests on its laurels that it is already a well-established science and everyone knows it is there if they would like to pursue it. In this day and age, when life is fast moving and the youth are into activities that keep them excited, we need to increase marketing. For example, we should hire whoever designed the milk campaign to promote the chemical sciences, because everyone is excited about drinking milk.

I think the youth would respond if they saw that going into the sciences could be exciting and create a future for them. It is left for people to know that if they want to go into chemistry, there is going to be a future in it. A lot of students don't see that, and they get steered into other disciplines.

Why is that not an option? I am going to put that question out to see if the panel would like to address it. Why can't we get some aggressive campaigning going to market chemical sciences—maybe all the way back through high school—and make them interesting, so that students will be excited and look forward to coming into these areas.

R. Stephen Berry: I will take that on first, just to cause trouble. I am not a proselytizer. My belief is that I want people to go into chemistry because they see it as a calling. I don't like to feel that I am persuading people that chemistry is where they should be. If they decide that, based on what they have experienced as undergraduates, that is great, and I try to make my undergraduate courses interesting. There are so many wonderful things to do in the world that I am not convinced that everybody should be subjected to my persuasion to become a chemist. I know that is anathema in this group.

Edel Wasserman: Just to keep this issue going, whether we advertise or not is not the main issue. I think the real issue is, Do we think we are here to fulfill the curiosity of people who want to—I will make it extreme—play in the chemical sciences, or is there a social purpose? I believe it is the latter. We are given money by the taxpayers to find the best people, who will be educated, go out into the world, and play a constructive role by contributing their chemical knowledge to society. If we can find them by getting the information out to the people so that they can decide that there is an opportunity that matches their talent and interests it is important to do so.

Dady Dadyburjor, West Virginia University: I would like to switch gears and return to the question we raised earlier regarding interdisciplinary work versus core competencies. If the question is interdisciplinary or core, I think the only answer to that has to be yes to both, in that each one nourishes and reestablishes the other. I have to disagree a little with Professor Berry. I don't think one does the interdisciplinary work just to reinvigorate a mature profession. I didn't think you meant that, but I wanted to say that I think each one builds on the other and reinforces the other.

R. Stephen Berry: I picked those examples as cases where it happened to work in that direction. It could go in the other direction as well.

Stanley I. Sandler, University of Delaware: I am a chemical engineer, not a chemist, and therefore perhaps I see things a little differently. Only a small percentage of our chemical engineering graduates go into academia.

I am troubled by this discussion of the generalist versus the specialist. Ron Breslow commented that a doctor still needs only 4 years of medical school, but in fact, medical practice has changed, as I discovered recently when I needed orthopedic surgery. I found that there are surgeons who specialize in knees, others who do only hands and fingers, or only shoulders, and so on. In general, if we look at other professions, whether medicine or law or even tax accounting, we find they have all become very specialized.

The question then is, What should our role as educators be and what sort of education are we trying to deliver? For future academics, the example is clear: being a specialist is what it takes to get tenure. However, from the industrial point of view, as we have heard here, a generalist seems to be more appreciated. I don't see how we can do both, so how should we decide on the appropriate goal?

Also, our discussion so far has been centered on the Ph.D. and maybe the postdoc as being the termination of formal education. With the present situation of frequent (and sometimes unplanned) career changes, and also because of the explosion of new technologies and scientific advances, there would seem to be a role for formal continuing or lifelong education. Indeed, this could become an increasingly important function of academia in the future. However, there has been no discussion of this, even though it is becoming more and more common in professions like medicine or law. In contrast, formal continuing education seems to play only a very minor role in the sciences and engineering.

Catherine Fenselau, University of Maryland, College Park: I am fairly familiar with the departments in a number of universities in the Baltimore and Washington, D.C., area. I would suggest that there has been a sincere effort to respond to the 1995 NSF discussion and the publications and suggestions that followed from it. I think that there is a great deal of interest across the country in issues associated with graduate education and a real recognition of the problems and opportunities. Many departments would like to see some strong national leadership. When we discussed some curriculum reforms recently at College Park, Maryland, someone said, "What exactly is it that ACS wants us to do?" Consequently, I would like to second the suggestion that was made, that perhaps a best-practices document could be put out by the new ACS graduate office.

Laren Tolbert, Georgia Institute of Technology: I would like to return to an unrepresented group here, that has been mentioned from time to time, which I think is an important part of the discussion, and that is taxpayers. Speaking as a liberal Democrat, I find myself uneasy in this role, but in fact, what we are talking about is going to be financed largely by the public. It does little good to talk about what the new programs are going to be if there are not funds to make that happen. If you look at what the taxpayers and the Congress respond to, it is initiatives in fields such as cancer research. Those are heavily research-oriented activities. So, we can talk all we want about improving graduate education but, in fact, the research model that gets the money is the one that is targeted toward specific research agenda. We need to take that into consideration, and that certainly should affect what we do. Otherwise, all we are going to have are these relatively small programs such as IGERT, which are not going to have a major impact on graduate education.

Derrick Tabor, National Institute of General Medical Sciences (NIGMS): I would like to address the question of whether or not diversity is a valid topic for this group to consider. Our group, Minority Opportunities in Research, is dedicated to increasing the number of underrepresented minorities in the sciences. So, this is a very important issue to us. Some of you may have seen the article, I think it was in the *New York Times*, that showed 100 Ph.D. degrees were offered to African Americans. I ask this group in particular, Who is responsible for that? This number could certainly be higher. It does not include Native Americans or Hispanic Americans and was not broken down in terms of women.

I think that this group needs to be asking, What more can we do to make sure that science is open to everybody, and not just to a specific few. So, I definitely would like to bring that question up again and to say that NIGMS is interested in knowing what you can do and would be interested in doing. Our director, Marvin Cassman, has addressed this in his guidelines, in terms of what is important for NIGMS to do. What can we do to encourage more creativity among educational institutions and graduate institutions to address this question? You are some of the most creative people in the world, but this seems to be one of the most difficult questions to address.

The second thing is that the MIRT program, which is Minority International Research Training, addresses globalization. It was started by David Ruffin's office to make sure that minority students, underrepresented minorities, are exposed to issues in international science. We send students all over the world. I took a couple of students to Australia, and we send students there every year. So, there is an active program. There are opportunities, and we need to support them, encourage them, and recognize them. This program is for students at the undergraduate level. So, when they get to your institutions, some of them may have already gone away for an international experience.

Edel Wasserman: I would like to comment on the last speaker's remarks. I think we don't have hard evidence as to whether increasing diversity will affect the fundamentals of science, although it clearly will help the world we live in. I don't know if there is something unique that a particular minority group or gender group has to contribute to science. Are there fundamental changes or fundamental things that are done only by women, and not by men, in the laboratory? Perhaps. If so, we ought to find these things out. By and large, I think it is a secondary question. These are major groups in society that we have to work at bringing into the graduate education process.

Derrick Tabor: Let me say this. I want to applaud the ACS for having the forethought to have the Minority Scholars Program. I think that the SEED program is fantastic. I am rather perplexed as to why you would say what you just said regarding minorities. My greatest fear—and I have not before expressed this publicly—is not as a National Institutes of Health (NIH) employee but as a member of the ACS. After those minority students complete graduate school, there will be no place for them, and this is not something that ACS has been planning for. In other words, these students will graduate and become members of the ACS, but there will be nobody out there to say, "Come to my table. We have been waiting for you because we believe diversity is important."

Edel Wasserman: Let me respond briefly to that. I don't know what the ACS will do with it, but please have the students contact me at DuPont. There will be a place at the table.

Isiah Warner: When you look at the composition of advisory boards for various companies in recent years, you will note that they have brought in women and minorities to serve on these boards. Suddenly they are recognizing that a diverse board brings a different perspective to the table. That is also true of science. You approach science from your own cultural perspective. I don't mean that a diverse group changes the basic tenets of science, since those tenets are absolute. However, one can vary in terms of how you bring others into science or how you teach science. I suspect that is why, and not because I am black, I attract a lot of black graduate students. It is because culturally I look at science somewhat differently. I think all of that contributes to the betterment of science in this country and that is the aspect on which we need to focus.

Lynda Jordan, North Carolina Agricultural and Technical State University: As someone who is a woman and of African descent, I am aware of the contributions of both minority and women scientists. I would remind you of the contributions that we have already made to the chemical sciences. Look at George Washington Carver, who at the early part of this century made significant contributions to society, and more recently Henry Hill, who is annually recognized by the ACS at the National Organization of Black Chemists and Chemical Engineers meeting. These are only two of the many contributions of minority scientists in this country and the world to the advancement of science.

The mere fact that we have to address these issues at a meeting of this caliber indicates the major biases that are associated with diversifying the demographics of the chemical sciences in this country.

4

External Research Collaborations Enrich Graduate Education

Lynn W. Jelinski
Louisiana State University

To prepare for this talk, I had two choices. The first was to give you a somewhat pedantic and dry discussion of my view of the value to graduate education of external collaborations. Or, I could go out and interview students and their professors, hear what they had to say, and share that with you. I did the latter. I set out to visit chemistry graduate students in their labs at Louisiana State University (LSU) with my digital camera in tow, and with a notepad and pen in hand. Once I arrived in the chemistry building, I asked selected graduate students what they thought was beneficial and enriching about their education and their external collaborations. Then I spoke with their professors. I was surprised at what I learned, and I think you will be, too. You realize, of course, that this study is not very scientific, because I interviewed only a few students and they were the ones I could get to easily.

Before I share my findings with you, I would like to find out whether we have some preconceived notions about the benefits of external research collaborations. I want to hear what you think is beneficial to the students, and at the end of my talk we will compare our ideas with what the students had to say (see Box 4.1).

Now, I would like to share with you what I learned from the students. There are three case histories I want to tell you about—I've changed their names to protect the innocent—and then I want to talk about professors who are going beyond the usual. The first student is Amelia. She gained confidence by doing something she had never done before.

The second, Arby, learned to love hard work. In fact, when Arby signed on with his advisor, he said, "You'll need to kick me in the butt because I'm lazy." Then there is Antonio, who learned how to give and take. Finally, I want to share with you the vision and dedication of some special faculty members who go beyond the usual.

CASE HISTORY 1: AMELIA

Let's start with Amelia. She is a nontraditional student who didn't go directly to graduate school. She has a child to support, and she is now eking out a living as a student and making the best of her

BOX 4.1
What the Chemical Sciences Roundtable Participants Named as Benefits for Students Engaged in External Research Collaborations

- Challenges
- Friendship
- Job opportunities
- Access to equipment
- Mentorship
- Travel, conferences
- Money

graduate education. It was obvious, both from talking with Amelia and with her advisor, that she lacked self-confidence. Fortunately, her advisor gave her the opportunity to collaborate with a chemical company, and it provided her with a chance to shine and gain tremendous self-esteem and self-confidence.

Amelia was tasked with delivering some high-quality nuclear magnetic resonance (NMR) spectra for a polymer company. The project sounds easy, doesn't it? Well, it wasn't. Somehow the spectra never turned out the same from day to day, and samples that should have nearly identical spectra afforded wildly different ones. The spectra were not reproducible. Amelia didn't trust herself to begin with, and now there was blame and subsequent squabbles among the folks in the lab who were in charge of the spectrometer, and even with the folks in industry, as they tried to figure out what was going on. Many issues had to be resolved, the least of which was obtaining sound and reproducible spectra. To make a long story short, Amelia discovered that the compounds were undergoing oxidation. If you have ever done NMR, you know that oxidation will definitely affect the reproducibility of spectra! Amelia independently figured out what was happening, and she became a hero in the lab and to the company.

But more to the point, this experience is a beautiful example of the ability to gain self-confidence. The experience of gaining self-esteem and self-confidence was echoed earlier today by someone in the audience who was talking about the value of a graduate education and said that part of it was "gee, look at me, I did it."

I asked Amelia to tell me what else she gained by working with industry. She said that the most important thing she learned was how to communicate well. She said that she learned not to work through a liaison, if at all possible, but to work instead with the bench chemist. She also learned how to get along with co-workers. "This is a lesson for life," she said, "as we love to blame everybody but ourselves."

CASE HISTORY 2: ARBY

As I mentioned, Arby is the graduate student who warned his advisor: "I'm lazy." According to his advisor, Arby is a very gifted synthetic chemist, with special expertise in organophosphorus chemistry.

BOX 4.2
Arby: Did Your Graduate Education Prepare You for Your Industrial Career?

Yes

- Fortunate to be doing exactly the same work as in graduate school

No

- Learned that teamwork is absolutely essential
- Not prepared for industrial scale
- Learned that, in industry, you don't have the luxury of getting two more data points
- "Fun, but I've been working 12-hour days!"

Arby's external collaboration project was well suited to him, as he was to perform an organometallic synthesis for a small, local chemical company. Over the summer, he was to alkylate some phosphines selectively for the company, alkylations that had never been performed before.

What is exciting about Arby's experience is that his summer project evolved into something more permanent. His industrial mentors recognized Arby's incredible talents, and his project quickly became a real job, and an important one at that. While working on his graduate education, Arby is simultaneously the lead chemist—these are his words, *lead chemist*—at the company. He was offered a senior position, and he is very happy with it.

Arby admits that he didn't work hard enough when he was a graduate student without an industrial job on the side. He produced no papers in his early years as a graduate student. In industry, however, he notices that there is a very different measure of productivity. Arby says, and this is a direct quote: "I am working harder now than I have in my entire life, and I am loving it." Arby also admits that he is paid a lot, and is proud of that.

I asked Arby how well his graduate education had prepared him for his industrial job, and here is what he had to say (see Box 4.2). I must admit that I do not agree with the only entry in Arby's "yes" column. He said he was fortunate to be doing exactly the same work as a graduate student as in the company. I thought to myself that the last thing you want is to do the same thing over and over. Right now, though, he feels that in terms of education and technical skills, his graduate education did a great job preparing him.

The answers in his "no" column are more interesting. It took his industrial experience to learn that teamwork is essential. We have heard from two students so far, and both emphasize the value of learning about teamwork. He also was not prepared for the industrial scale. The day I talked to him, he had just come back from a hazardous operations briefing about a reaction he is scaling up to 2,000 gallons. In industry he does not feel that he has the luxury of getting two extra data points. I suppose we all know what that is like. Finally, I think Arby understands what hard work is all about—and he loves it—and that is something that he did not appreciate at the start of his experience.

CASE HISTORY 3: ANTONIO

Antonio learned a tremendous amount about give-and-take. His project involved providing some bromine nuclear quadruple resonance (NQR) spectra for a large company. The project required looking at flame retardants in high-impact polystyrene.

This was a very tough project, as bromine NQR spectra are notoriously difficult to obtain. In fact, one of my colleagues at another university is quoted as saying,"If you have a death wish, try bromine NQR." But the value to the company of the information to be learned was worth the difficulty, and Antonio eventually produced beautiful results.

Antonio learned new skills along the way, skills that we perhaps take for granted. For example, he learned to think about research, he said, in a logical and clear way. He says that learning this has helped him in research and has also carried over to his whole life. He said he learned to give and take, he learned teamwork—we have now heard this from all three students—and he learned how to *listen*. He recognized that this is also a lesson for life.

Finally—and Ed Chandross, this gets back to your comment on how industrial collaborations can lead to publications—Antonio learned that industrial people, too, go to meetings and publish. The industrial collaboration led to a joint publication in *Macromolecules* and to a joint poster presentation at the Experimental NMR Conference (ENC). In fact, when I was visiting Antonio in his lab, he was so proud of his collaborations that he gave me one of the reprints that just came out and showed me the poster on the wall.

OTHER ISSUES

The other thing I would like to point out is that the professors also learned from these external research interactions. They, too, learned about working with people and students and balancing both sides of the equation.

So far, I have discussed external collaborations via case histories. These were just snapshots taken through the eyes of students who were working, for the most part, on short projects with specific deliverables.

If we put our administrative hats on and think about students working at industrial sites, there are other issues that need to be addressed (see Box 4.3). There is, of course, the issue of intellectual property. Most universities have resolved this fairly well. There is also the issue of communication. For example, are we really on the same page? Who is running the show? Is it the professor and the research director, or is it the student and the bench chemist? Who is driving the research agenda? The

BOX 4.3
Other Issues Pertaining to External Research Experiences for Graduate Students

- Intellectual property
- Communication and who runs the show
- Working on projects related to professor-owned companies
- Sharing of information

BOX 4.4
What Three Graduate Students Identified as the Value of External Research Collaborations

- Learning the value of teamwork and communication (all three students)
- Self-confidence through addressing *challenges* (all three students)
- Learning how to get along with co-workers (two students)
- Learning how to work hard
- Learning how to think logically
- Publications
- *Conferences, travel*
- *Money*
- *Job opportunities*

Note: Italicized entries are those that were mentioned by both the roundtable participants and the students.

last thing to think about is an issue that we think we have resolved at LSU, but it is important to address. How do we handle students who work on projects of professor-owned companies? Because more and more professors share in ownership of companies, we need to be careful that we do not exploit graduate students who might work on projects related to those companies. There are also some real ethical issues having to do with what you share in terms of information. I hope that some of these will come up in our discussion.

Let's take a look at the expectations of what the audience thought, at the outset of my talk, about the value of external collaborations, and compare those thoughts with what we learned from the students. As Box 4.4 shows, we selected some of the attributes (the common ones are underlined) that the students also thought were important. But we, the roundtable discussants, missed several values that were echoed repeatedly by the students. We missed the value of learning teamwork and cooperation. And, to some extent, we underestimated the value of gaining self-esteem and self-confidence by successfully addressing challenging problems.

PROFESSORS: BEYOND THE USUAL

The last thing I want to discuss is what I call professors who go beyond the usual. The outside collaborations I shared with you had to do with industry, but there are many other kinds of outside collaborations.

Some of our professors, led by George Stanley, are dedicated to making K-12 outreach a part of the graduate experience at LSU. If we could get *all* of our graduate students—not just those in chemistry and chemical engineering—to spend a few hours a month working with local students in our K-12 schools, we would go a long way in bringing K-12 students to where they need to be. Could you imagine the impact if this were replicated across the nation? If we could have other universities like

LSU take part in this teaching, work with the kids, and excite them about a favorite subject, we could make a tremendous difference.

Isiah Warner is another example of a professor who goes beyond the usual. Isiah has been legendary as a mentor. He won the Presidential Mentoring Award and a few weeks ago received the Lifetime Mentoring Award of the American Association for the Advancement of Science. And, he has added yet another award to his list, this time from the Eastern Analytical Symposium.

Professors like George Stanley and Isiah Warner are why I truly do not believe that there will be an effective Internet university of the universe. In the end, everything boils down to *people*. The point is that there are magical people, and all of us know who they are in our universities. We know who those magical people are who work with the students, who encourage them, and who mold them into being the best that they can be.

DISCUSSION

Victor Vandell, Louisiana State University: One thing that you are pointing out, and I think it is something that we noticed when you came to LSU, was your concern about people and relationships. You value and obviously place importance on individuals, especially in a mentoring capacity, and the impact that they can have on the students. I think we see that in your presentation as well.

Ernest L. Eliel, University of North Carolina: You obviously have a cooperative program. We talked earlier about the time factor in graduate study. How does that work out? I mean, how much time does the student spend in the industry? Is that part of his or her thesis? How does that affect the total time spent in graduate study?

Lynn Jelinski: It varies. Much of what I showed you today is simple—running a spectrum here, running a spectrum there, but very short projects in general.

Anne Duffy, University of California, San Diego: I am a graduate student. If I had to say what the most important thing is for me, I would say it is mentoring. I went through an extensive interview process when I was looking at graduate schools, because I wanted to make sure that I would be able to work with a compatible faculty member. While doing that, I discovered that one of the most compatible people was at the place where I had obtained my undergraduate degree, so I stayed and worked with her a little bit before I went on. As far as industrial relationships are concerned, I also think they are important for people who want to go on to industry.

I worked before I went to graduate school. I guess I am what is called a nontraditional student because I have children and am married. I have been a supervisor, have hired and fired, and know what it is like to work on a team. I see many graduate students coming in right out of college because they don't know what else to do. So, they go to gradate school, and keep going, until they get spit out into the system and have to decide what to do then.

Jodi Wesemann, St. Mary's College of California: As a faculty member, I have been encouraging internships for undergraduates at St. Mary's as a way to expose them to what is beyond college and help them make an informed decision regarding whether to go to graduate school or not. One of the things we have run up against, which I am sure is common at graduate and undergraduate institutions, is the time that it takes for the faculty to support and put together a program like this. My question for you is,

How much support exists outside the department, i.e., the infrastructure that is in place to arrange the internships, or do the faculty have to coordinate all of them?

Lynn Jelinski: I think the faculty must coordinate a lot of them.

Judson Haynes III, Procter & Gamble: Before coming to Procter & Gamble, I completed my Ph.D. at LSU with Dr. Isiah Warner. I want to share one of my experiences in graduate school that was really a turning point to me. It involved basically my first project. I began working on it and, after two years of struggling, I had some data to write up in a paper format and submitted it. It was returned by the reviewers with some not-so-nice comments. I went to Dr. Warner and said, "Dr. Warner, I don't like this, this is scary. This is two years of work." But the most important thing, and what I call a turning point, was that Dr. Warner asked me what my grievances were. I wrote them down and we addressed the editor's comments and got the paper published. We were able to find literature references to previously published work. It was incredible. If he had just sat back and said, "you have to do what the editors say," I don't think I would be in chemistry. I would have probably said, "No, this is 2 years of my life that I spent in a lab, and somebody is telling me that it is not good enough." That is a very important part of graduate education, to reinforce to the students that you are there to support them and to help them grow.

John Schwab, National Institute of General Medical Sciences: Today we have been hearing all sorts of wonderful anecdotes about the importance of mentoring and about the impact that really effective mentoring can have. My question is this: How do we go about institutionalizing this within the entire field of chemistry? I would propose that what we have right here is a chosen group. We are essentially preaching to the converted. I think, and I believe almost everybody here would agree, that this is something for which we need to have disciplinewide standards, and we need to have some way of enforcing standards of mentoring. I would like to open the floor to responses to that challenge.

Lynn Jelinski: Good. As I said, I certainly don't have all the answers. So, I am going to ask the audience to respond through this rapid-fire question-and-answer session. Does anybody want to respond to the question?

Brent Koplitz, Tulane University: I interact with the Louisiana public school system. We have tried to address some of the issues associated with this problem through what we call outreach, a concept many of you also employ. One of the things I would like to put forth as a challenge to the national funding agencies is that over time we can create a new graduate category. We have TAs, teaching assistants; we have RAs, research assistants; and we should have OAs, outreach assistants. I put that to the people here as something that would help chemistry, not only with funding, but also with its exposure to elementary, middle, and high schools. A program like this is especially needed in urban areas; it is something whose time has come. It will do a lot of good public relations. I have put this in a proposal before, but unfortunately it was not funded. The proposal has, however, allowed me to coin that term. There are plenty of graduate students who are interested in a program such as this one, and they are the ones to do it.

Peter Eisenberger, Columbia University: I have a similar question on the external experience or industrial experience. It seems that it should be almost a necessary part of graduate education to have an alternative culture that you are exposed to besides the academic culture. The question I have is, How do you institutionalize that? Some people—I know that Bob Dynes [Chancellor of the University of California, San Diego] is one—have forced every one of their students to spend at least one summer in industry. You are asking students to make a life decision without anything to compare it to if they do not have exposure to another culture.

Lynn Jelinski: This comes back to your point about institutionalizing. Are there any other students that we haven't heard from?

Jonathan Bundy, University of Maryland, College Park: As a graduate student, I think it is important that this mentoring start at the undergraduate level. We need to get undergraduates in research groups, get them involved with chemistry in the raw, so to speak. I think in that way we might have more American students in our graduate programs.

Lynn Jelinski: That is a really good point.

Jonathan Bundy: Not just the best and the brightest.

David Budil, Northeastern University: I don't know how many people are aware of it, but Northeastern, at least, as an undergraduate institution has made its reputation as being a co-op school. Undergraduates spend a year of their 5-year undergraduate experience working in the field. Graduate programs have tried to emulate this. It hasn't been as successful as it should be in chemistry. I think part of the reason is that our internship program has been subject to the ravages of economic cycles. I am going to throw out a question I hope some people would address. How do you protect an arrangement like that, once it is in place, from falling victim to inevitable economic cycling?

Lynn Jelinski: Are there any financial gurus in the audience? It is a really good point.

Carolyn Ribes, The Dow Chemical Company: I want to emphasize that the things the students thought were important are the same things that we at Dow think are important. When I am evaluated on how I do my job, it is not just my technical knowledge and my problem-solving abilities that are considered, but it is also on how I do in areas of initiative, leadership (as mentioned by one of the students), teamwork, and interpersonal effectiveness. All of those skills are going to be measured in your career. All of those things are valuable. A lot of times we focus only on the technical issues when educating students.

Lynn Jelinski: Great, I wish all of our students could hear you.

Timothy A. Keiderling, University of Illinois at Chicago: I would like to make a quick comment. The NSF did support 31 outreach-oriented grants this year with graduate assistantships to go into schools. The University of Illinois at Chicago has one of them. If you send your education-oriented students to us, we will give them fellowships and send them into the schools of the city of Chicago to try to solve real education problems.

Second, I think we skipped over it, but it is becoming a real issue that Peter addressed earlier. Universities are becoming more entrepreneurial. The bottom line is dollars. In response, faculty are forming companies. We have research parks on both of our campuses at the University of Illinois. The idea is to put incubators there, to have faculty set up companies, to have these ideas develop into technology transfer, and the university gets a bigger share of potential profits. Students are involved with this, which is a real ethical issue with which we regularly contend. We have established oversight committees, but they do not work perfectly. I am interested in other people's methodology and thinking on this.

Lynn Jelinski: We have solved this by putting in place a local oversight committee. What you want are faculty colleagues in your backyard watching you. Maybe we could hear from others.

Isiah Warner, Louisiana State University: What I want to comment on is that much of industry tends to be like academics, i.e., elitists. For example, many of you know that in Louisiana we have 300 chemical companies along the Mississippi River. How many companies spend time interviewing our students at LSU? Not very many. Most of these companies have their headquarters somewhere else. Dow Chemical is one example. Dow rarely recruited from LSU until recently.

They interviewed a few of our students at the American Chemical Society meeting, which happened to be in New Orleans (our backyard). The head recruiter from Dow called down to Dow at Plaquemine, our next door neighbor, and said, "What are you guys doing? Why aren't you recruiting at LSU." The local recruiters basically said, "Well, you guys have always told us where to recruit. We only interview at the top 10 schools." All of this conversation transpired because headquarters interviewed Victor Vandell and three or four of our other students and found that we have talented students at LSU. Indeed, they were very impressed. Again, I think that part of the problem is that too often we in academics and industry tend to be too elitist.

We need to identify a broad spectrum of institutions that are training different kinds of students in a way that is compatible with the kinds of students you want to recruit to industry. I think that is the most important criterion, not that they come from the top 10 schools but that you are getting the kind of personnel that you need in your laboratories. Suddenly, Dow recognized this after they interviewed a few of our students. We are now on the recruiting list of Dow. How many other companies would do the same thing if they would recognize the talents of our students?

Ellen Fisher, Colorado State University: I am going to bring up a negative side of this, which is that I have had a couple of interactions with industry, including funding through one of these consortia that was spoken so highly of this morning. My experience and my students' experience have not been as positive as the students that you interviewed. We have had issues with members of this consortium bickering back and forth, being distrustful of our results, and questioning the validity of them. They then suggested, and in fact someone told me, that we had not done the definitive experiment. When I asked what the definitive experiment was so that we could do it and prove that these were accurate results, he was unable to come up with an experiment that would prove this. What my students have seen from interacting with industrial people has therefore not been quite as positive. Some of the positive things they have gotten, but they have also come away with a bad taste, because they have not had a really good experience. I didn't hear any negative comments from your students. I was wondering if they had any.

Lynn Jelinski: Mine was a completely nonscientific sampling. Although none of the students I spoke about today had a bad experience, I am sure some out there have difficult experiences.

P. Wyn Jennings, National Science Foundation: I want to tell you that, in fact, there is a program in place at NSF for taking graduate students and some undergraduates into K-12 education. That program was in its inaugural year last year and will continue as the budget is increasing. It is a traineeship, taking 10 to 20 students a year. The program is called GK-12, and it is on the NSF Web site at <http://www.ehr.nsf.gov/EHR/DGE/GK-12.htm>, if you are interested.

Ronald T. Borchardt, University of Kansas: I guess that I am an outsider here, since I represent the biological and pharmaceutical sciences. However, I wanted to follow up on what John Schwab mentioned. That is, many of the things that you have been talking about here today—interdisciplinary training, internships, mentoring—have been significantly influenced in the biological and pharmaceutical sciences over the last 25 years by the National Institutes of General Medical Sciences (NIGMS) through its training grant programs.

These training grant programs have had a significant influence on the way many of us in the biological and pharmaceutical sciences are now educating graduate students. It was not the NIGMS administrators who dictated change to us. It was our peers who were reviewing our training grant applications who told us that we needed to change the way we trained graduate students. I think that somehow chemistry departments have to develop similar review mechanisms having significant external input and influences (i.e., funding).

Lynn Jelinski: One way we've addressed this issue, that goes to exactly what you are saying is, "No department can get institutional matching funds unless they have an external advisory board."

Billy Joe Evans, University of Michigan: I would like to recall some of the things that Steve Berry mentioned, the idea of flexibility. In flexibility, there is a dominating concern for authenticity. If there is no authentic need in the student's work to go into industry, why must he or she do it? If there is a need, then it should be done. If there is no need, then we shouldn't encourage it, because you are liable to create something you did not intend.

Just as Steve Berry mentioned that Linus Pauling did in his research 36 years ago, when I did my course work in chemistry I did my thesis work with a geophysicist. I worked in low-temperature laboratories, and I used the CDC 3600 computer at Argonne, because the computer at the University of Chicago was not good enough to do the work we were doing. In the process, I also worked with nuclear physicists at Argonne Labs. I must have published four or five papers with them in addition to the work that I was doing at Chicago. I finished in 4 years, and I had seven or eight papers when I finished. But those were authentic connections. I needed those.

My advisor was fully in control of what was going on. I had one mentor. I was bound—if anyone knows what that means—I was bound with one person. Steve Berry was a good stimulus, and J. Willard Stout was a stimulus to me, as was Julian Goldsmith and Stan Ruby at Argonne. But I was bound with one person. Institutions do not mentor. Institutions can have environments that promote that, but institutions do not mentor. As your student mentioned, people mentor. One can't have too many mentors at one time.

Craig A. Merlic, University of California, Los Angeles: One of my concerns is that there are not always enough of these industrial internships to go around. Not every university is within the proximate vicinity of 300 chemical companies along the Mississippi River. So, there are other models that people can use. At UCLA, we have a chemistry and biology interface training grant. As one component of this training grant, graduate students are required to spend up to 6 months in another laboratory, working on different projects. I think a lot of the skills that they need, such as interpersonal skills and solving problems, can be learned in an academic lab. So, we can take care of the dearth of industrial internships by establishing these collaborations, which are not really collaborations, but are just internships in an academic lab.

Michael Doyle, Research Corporation and the University of Arizona: I was going to say about the same thing with regard to industry. There aren't enough industrial opportunities. What I would like to go back to is that the one individual that Dr. Jelinski said lacked confidence had an experience in which the results were brought into question and finally solved the problem in a marvelous way. Unfortunately, I think that the experience that we expect, both at the graduate level and, increasingly, at the undergraduate level is that you find your own way. You are not really a full participant until you have proven yourself. That is a concern. We are throwing a lot of money into research, and we are saying a lot about its role as education. But the process of how you develop an answer to a problem is still an issue that we have to confront.

Victor Vandell: I would like to address the issue of how we can institutionalize some of the ideologies that are coming through in meetings like this. One suggestion is that organizations that have the power of influence, such as the ACS, take fundamental concepts like mentoring, outreach, and teamwork and promulgate those ideologies to the individuals who are their members and to universities. If they devoted their time to promoting these ideas, these particular ideologies will eventually become a common theme in the institution and in the workplace.

Christopher J. Cramer, University of Minnesota: I have been a subcontractor on a National Institute of Standards and Technology Advanced Technology Program grant to Phillips Petroleum. I would like to point out that each October I waited around until Congress passed that program by about three votes, usually two months into the fiscal year, because it is "corporate welfare" and therefore not a good thing. For those policy makers in the audience, I would like to point out that there is a certain amount of dissonance at the highest levels about how good these industrial/academic collaborations are. It would be nice to have more harmony.

Thomas Edgar, University of Texas: In that light, I received a note from Steve Berry a while ago. He says that the National Academy of Sciences, through its Committee on Science, Engineering, and Public Policy, has published a report on mentoring called *On Being a Scientist: Responsible Conduct in Research,* Second Edition. It is available in full text on the Web via the National Academy Press <www.nap.edu>.

5

Portals to Knowledge: Information Technology, Research, and Training

Eric G. Jakobsson
University of Illinois at Urbana-Champaign

INFORMATION TECHNOLOGY WILL BLUR THE BOUNDARY BETWEEN RESEARCH AND EDUCATION

Let me tell you what I mean by "information technology," or the appropriate information environment for any field of study. What I mean is that you ought to be able to fire up your Web browser and have, at your fingertips, all the important information sources, plus the tools to analyze that information, calculational tools to correlate and make sense out of the raw data, visualization tools when that is appropriate, simulation tools, and molecular dynamics and quantum chemistry tools. It all ought to be available, point and click, on your Web browser. That is the kind of information environment that I am suggesting will blur the boundary between research and education.

If you create that environment, you will not only make the access to this information and the use of it more efficient for the research scientist, but you will also make it accessible to the student. You are going to incorporate this environment into the classroom, and students will use it for undergraduate research and for exploring and learning in a hands-on way, at least in terms of the computational sense of things, and in a hands-on way about the nature of their subject.

INFORMATION TECHNOLOGY WILL BLUR THE BOUNDARIES BETWEEN DISCIPLINES

Information technology is already blurring the boundaries between disciplines. We used to have students climb the ladder of knowledge. There was a ladder for chemistry, a ladder for biology, a ladder for physics, and so forth, and they were next to each other. Now, students navigate the web of knowledge. On the Web, you can move sideways as well as vertically. You can move in any direction as easily as any other direction.

If I want to know something about nitrogen, for example, and I look on the Web, it is equally easy for me to access a description of industrial processes for synthesizing ammonia, or a description of the natural processes by which nitrogen is cycled through the biosphere.

IT IS AS EASY TO MOVE ACROSS THE BOUNDARIES OF OUR TRADITIONAL DISCIPLINES AS IT IS TO MOVE WITHIN THE BOUNDARIES OF OUR TRADITIONAL DISCIPLINES

As more and more of our exploration is in knowledge spaces like this, the divisions between disciplines are going to be blurred. We have talked a little bit about specialization and generalization. It certainly is true that there is a long tradition of both interdisciplinary work and of scientists being generalists. The relative importance of interdisciplinary work has waxed and waned. It seems to me that after World War II there was a tremendous move toward specialization in science, which began with big federal funding of research. Perhaps this was the beginning of the various pressures that we have talked about in terms of being in the grant rat race.

I believe that information technology will start to reverse this trend; in fact, it already has started to make it possible for individuals to find out about things more efficiently, for individuals to become multiple experts. Indeed, because of the pace of change and the various tasks that people are going to do during their lifetimes, it will be almost mandatory for people to become multiple experts.

INFORMATION TECHNOLOGY WILL BLUR THE DISTINCTION BETWEEN HERE AND THERE

I certainly don't mean that interacting with people over the Web should replace mentoring. As collaboration tools get better and better, they are not a way for people to avoid communicating with each other, or avoid mentoring each other—they are other ways for people to communicate and mentor. I can tell you the innermost contents of my mind as well over the Internet as I can in person. In fact, I may be even less inhibited about doing so.

THE BIOLOGY WORKBENCH

I would like now to show you what I am talking about in terms of the Biology Workbench. Figure 5.1 shows what we call the National Computational Science Alliance (Alliance) Information Workbench, for which the Biology Workbench is the prototype. The user interface in this diagram is the Web browser. The Web browser connects to the guts of the Workbench, which is the Workbench server. The Workbench server translates formats, creates queries that databases can understand, and drives application programs. The application programs have various interfaces, as can the information sources, and can be written in different languages. All of these can be readily tied together by a powerful scripting language such as Practical Extraction and Report Language (PERL), which is the easiest to use. But you can script in C as well as in PERL.

The key is that we now have the ability to take, for example, a whole series of databases in varying formats, such as various molecular biology databases, and a whole series of application programs, such as programs to visualize molecules, align molecular sequences, and construct phylogenetic trees of relationships based on those sequences, and so forth, and make them all look like one program. You interface through them and point and click as though you were on a Macintosh.

I know that this is not a Macintosh, but the point is to make everything in the world look like one. This is what we are really about. I tend to think in terms of biology, but basically the challenge is the same for chemistry. In chemistry you have databases that give you physical properties of chemicals; you have databases that give you structures, and so forth. For each discipline, what you really want to

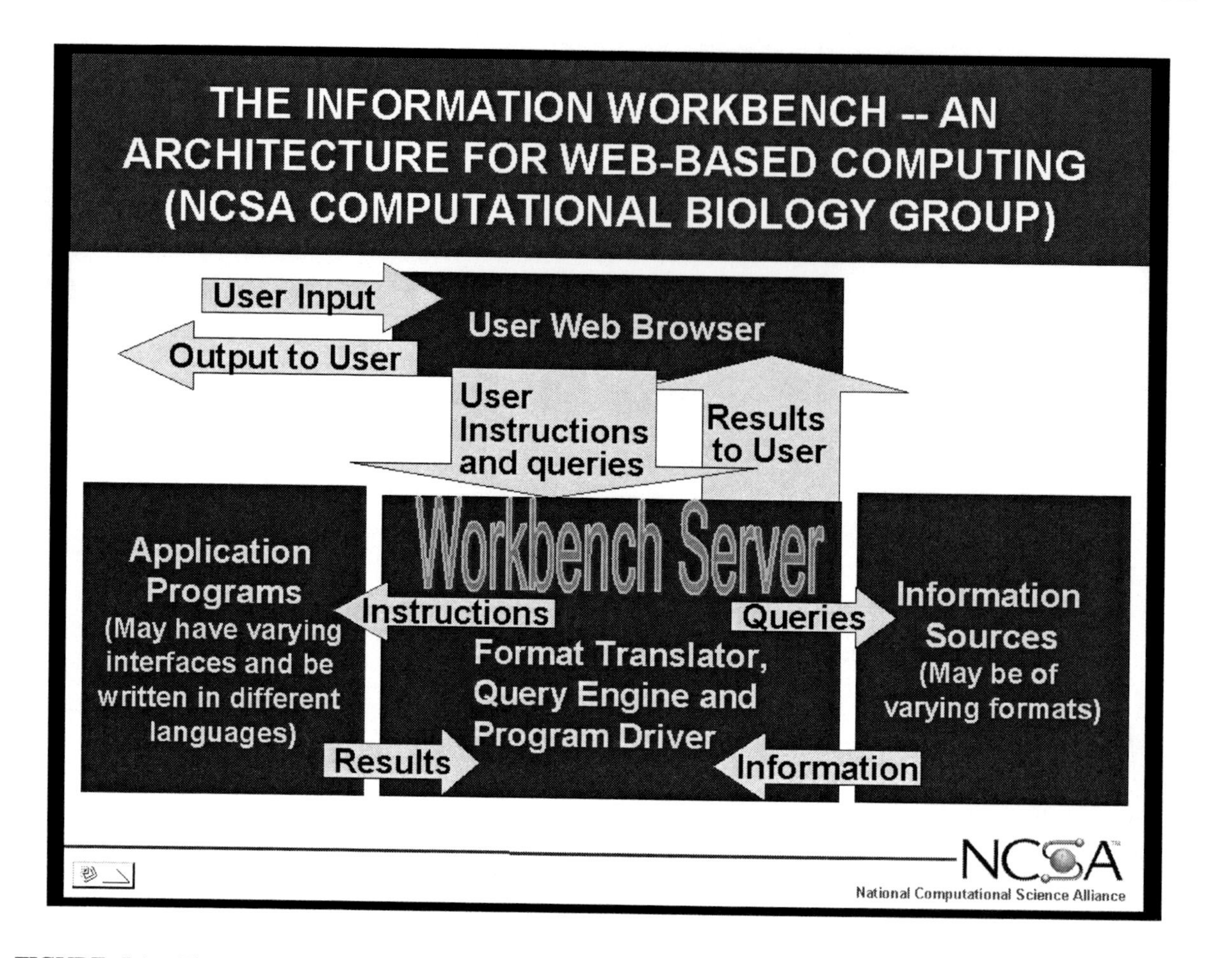

FIGURE 5.1 The National Computational Science Alliance Information Workbench, the prototype for the Biology Workbench.

have is a single, seamless computer interface providing access to all the data and visualization and analysis programs.

The next series of figures (Figures 5.2 through 5.12) are screen shots that show this type of interface for biology, the Biology Workbench, at <www.ncsa.uiuc.edu>. Figure 5.2 shows the interface for the Biology Workbench, with a picture of my colleague Shankar Subramaniam, the primary inventor and driving force for development of the Workbench. Although he has now moved to the University of California, San Diego, we still do some collaboration.

The Biology Workbench has multiple functionalities that can be accessed by simply highlighting and clicking on the desired function (Figure 5.3). To illustrate the capabilities of the Workbench, I will perform a search for human myoglobin. Once entered (Figure 5.4), the Workbench searches through multiple databases simultaneously, with thousands of sequences. The results returned by the search include the number of database entries found relating to "human," the number relating to "myoglobin," and the number that related to both (Figure 5.5).

One of the four databases returned from the human myoglobin search is illustrated in Figure 5.6. The database contains hyperlinks to such things as the original paper (Figure 5.7) and the structure

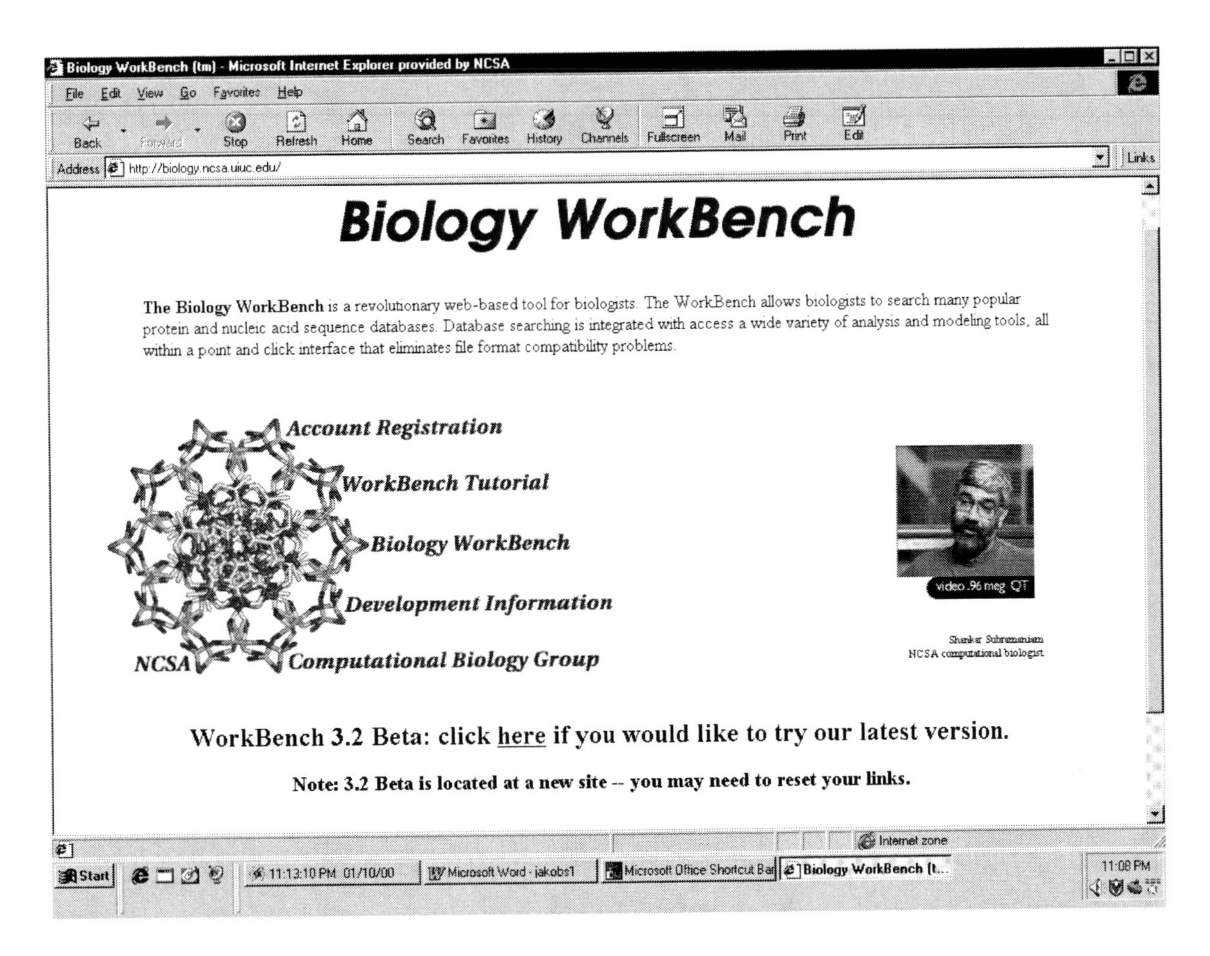

FIGURE 5.2 The Biology Workbench interface. Shankar Subramaniam, the primary inventor and driving force for development of the Workbench, is pictured on the right side.

(Figure 5.8). The structures can then be manipulated, and various measurements can be done on them with a plug-in called Chime, which is familiar to many chemists as well. It is really a computational chemistry tool.

The Biology Workbench can also be used to search the databases of amino acid sequences to find a particular sequence (Figure 5.9). The Workbench then allows you to take that sequence and use a program called BLAST to find closely related sequences. For human myoglobin, BLAST returned several "hits" of myoglobin sequences from a chimpanzee, another kind of ape, a gibbon, a gorilla, and an orangutan (Figure 5.10). The hit sequences can be aligned to provide a clear picture of sequence homogeneity (Figure 5.11). The regions that are highly conserved tell us one thing, and parts that are variable tell us another. The parts that are highly conserved must be critical for defining the basic structure and function of the molecule. The pattern of variability in those regions that are not conserved tells us the evolutionary history of the molecule and, by extension, of the organisms.

The Workbench can further be used to create a phylogenetic tree based on the pattern of differences in the amino acid sequence alignment (Figure 5.12). For our sample case, it is seen that the human (myhu) is grouped with the African apes (the chimpanzee, mycz, and the gorilla, mygo), and this cluster

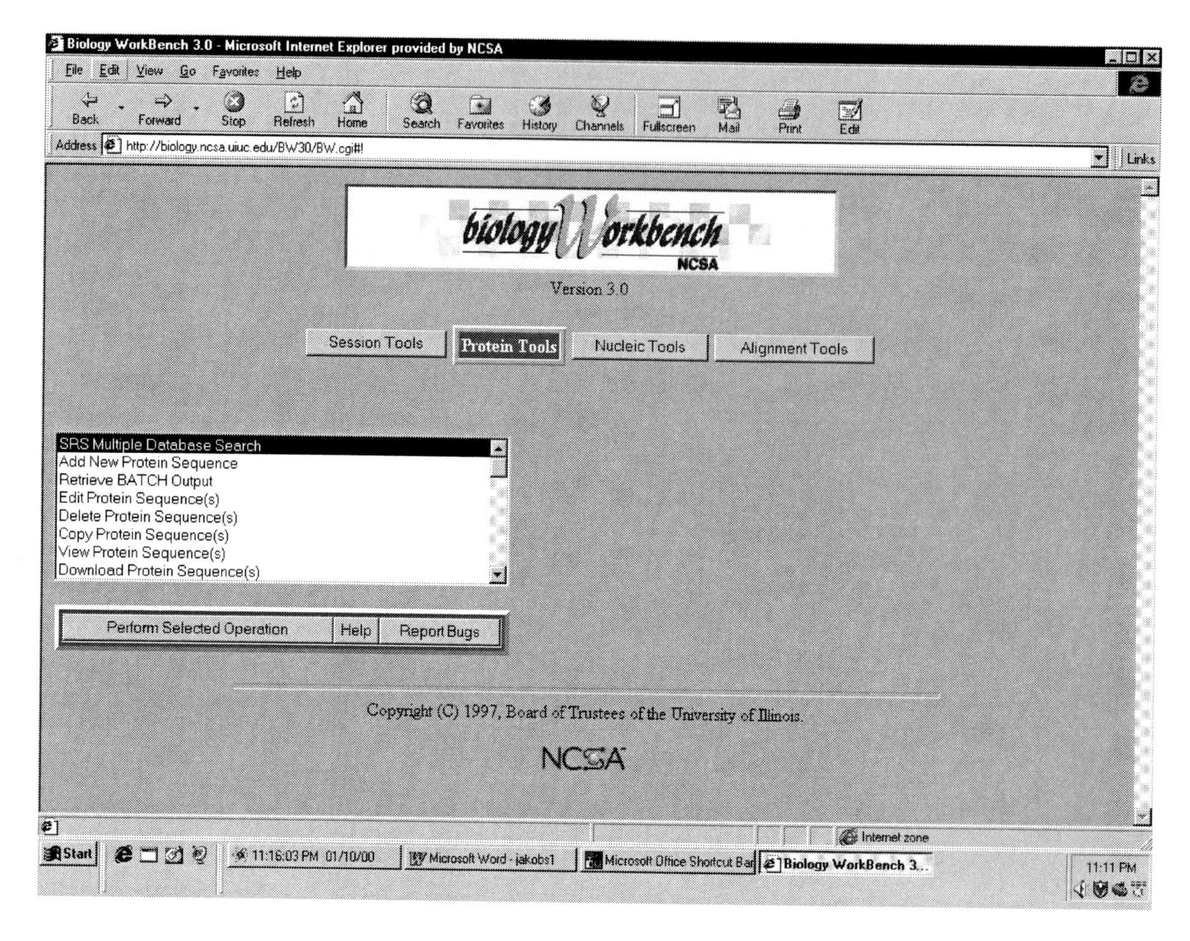

FIGURE 5.3 This picture illustrates the highlight-and-click process involved in selecting the functionality of a database search in the Biology Workbench.

is a short distance from the Asian ape (orangutan, myog), with the other primates farther away. This pattern is confirmed in the analysis of many genes.

All of the calculations shown in the above sequence can be done in point-and-click fashion in the Biology Workbench in 15 to 20 minutes. All the user needs is a networked desktop machine with a Web browser. The universal access, plus the ease and speed of accessing the data sources and doing the analysis, clearly makes this tool potentially useful in the classroom. But students and teachers need some introductory training in the use of the tool. Instead of trying, ourselves, to introduce the Biology Workbench to all of the faculty in the country, we started a process this past summer, in collaboration with the BioQUEST Curriculum Consortium at Beloit College, of building a community of biology instructors nationwide, working with them, and having them work with each other, to build curricular materials around the Biology Workbench and, in this fashion, to build a community of people who are building curricular materials.

We believe that the Workbench architecture is a prototype for the use of computing in education, and for much of scientific computing in general. It is the basis of the National Computational Science

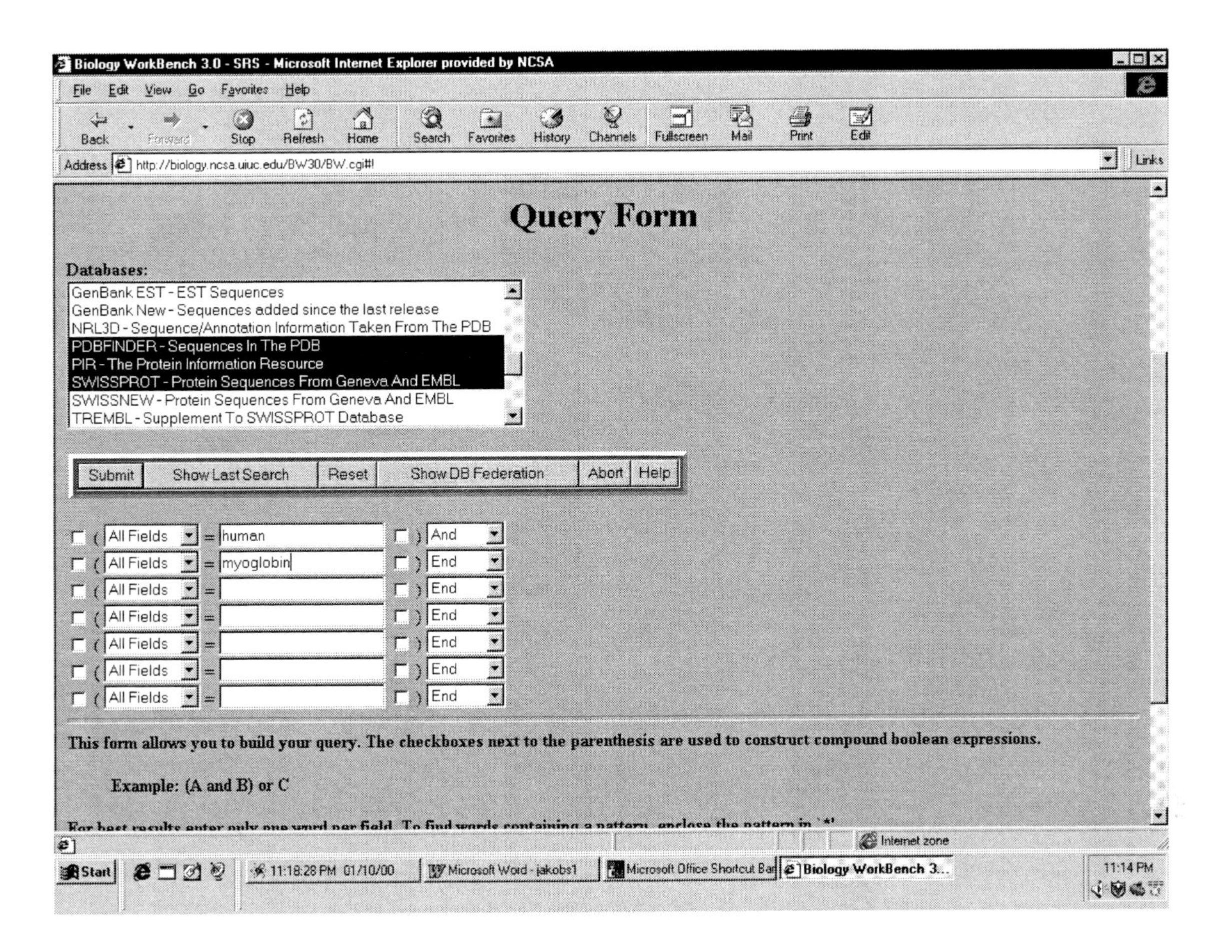

FIGURE 5.4 Screen from the Biology Workbench showing how to submit a search for human myoglobin to several databases at the same time.

Alliance Common Portal Architecture Project, which is summarized in Figure 5.13. A major part of further extensions of this kind of computer architecture is to build in collaboration tools so that people can work with each other remotely, and to extend the Workbench environment to more intensive computing than we now do on the Biology Workbench. The jobs that we do on the Biology Workbench are mostly not terribly computer intensive. It is reasonable for them to be done interactively.

We are currently extending the Biology Workbench capabilities to a variety of computational chemistry functionalities. We are building Web interfaces for molecular dynamics, stochastic dynamics, and electrostatic programs. These applications are computer intensive and will require us to do more extensive things with computing architecture to ensure that resources are set aside in advance for doing this type of work.

For many years it has been speculated that computers would have a big impact on education. Until very recently, this promise has not been fulfilled to a significant degree. Now, because of the extraordinary versatility of the Web for communication and data access, the enormous increase in underlying computer power, and the ease of use of the Web, we believe that high-performance, versatile, and user-

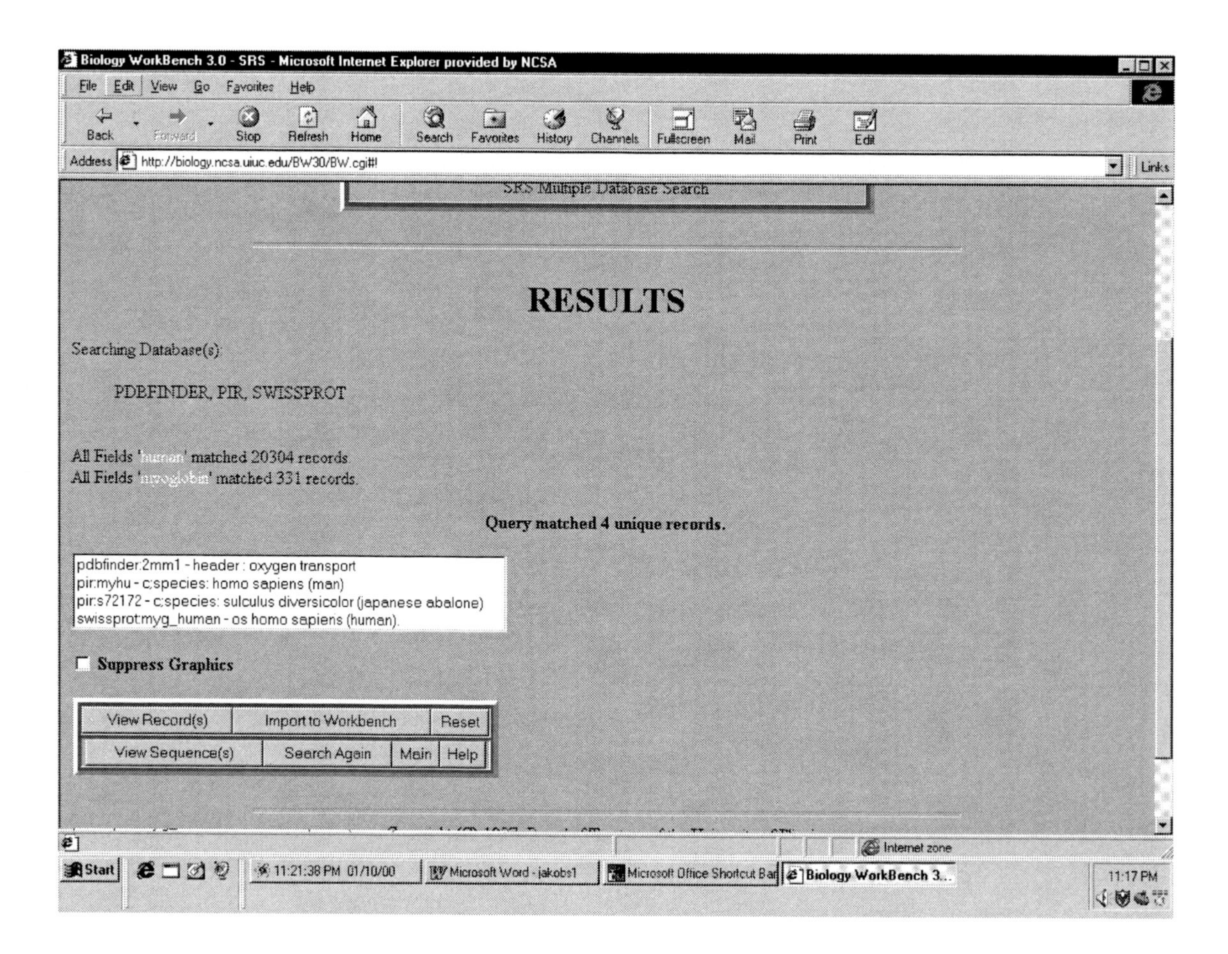

FIGURE 5.5 Results of the search for human myoglobin showing that it found 20,304 database entries related to "human" and 331 related to "myoglobin," with just 4 entries relating to both.

friendly computing and information environments are poised to have a profound impact on chemistry and all scientific education.

DISCUSSION

Robert L. Lichter, The Camille & Henry Dreyfus Foundation: A question that arises repeatedly, as we move into the technology of information access, is the value of the information obtained and the need for arbiters, referees, and other gatekeepers. What do you foresee as some of the challenges that arise in that context, as we proceed along this track, through all the portals, collaboratories, and the rest?

Eric Jakobsson: It is an enormous challenge, and there are of different ways to deal with it. For example, the physics community, which has a big repository of papers in Los Alamos, doesn't vet anything before it goes in and prints everybody's comments. So, if somebody publishes a paper about

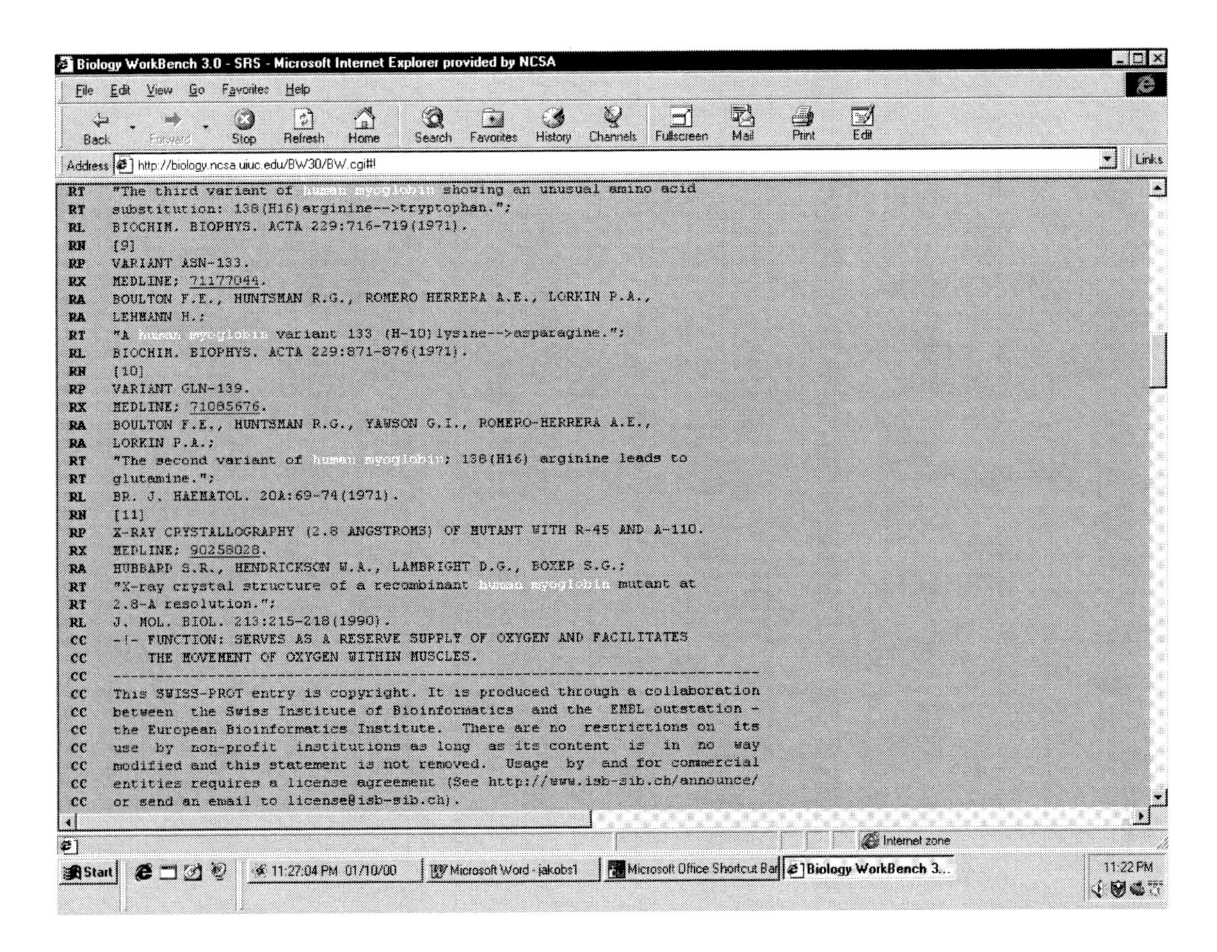

FIGURE 5.6 Part of one of the database entries given for human myoglobin. The entry includes hyperlinks to the original paper as it is indexed in the National Institutes of Health PubMed query page.

a perpetual motion machine, they count on other people heaping scorn upon it; the comments are attached to that paper, and nobody will pay any attention to it.

Biologists tend to be more conservative than that; they are very concerned about the accuracy of information. In fact, there is a growing amount of literature about people finding mistakes in the sequence databases.

My colleague Shankar Subramaniam has developed a very powerful algorithm called PDS, Potential Density Functions. It assigns parallel probabilities of interatomic distances between different kinds of atoms in a protein structure, based on statistics derived from known structures. It turns out that this is an extremely good filter for finding mistakes in structures in the protein data bank. As this data proliferates and more and more people depend on the data bank, we are going to need to pay more attention than we have to the accuracy of the data.

I should say that currently the big emphasis is on dealing with the volume of information. Mostly it is really good, but there is enough that is not that has become a nuisance, especially when you build

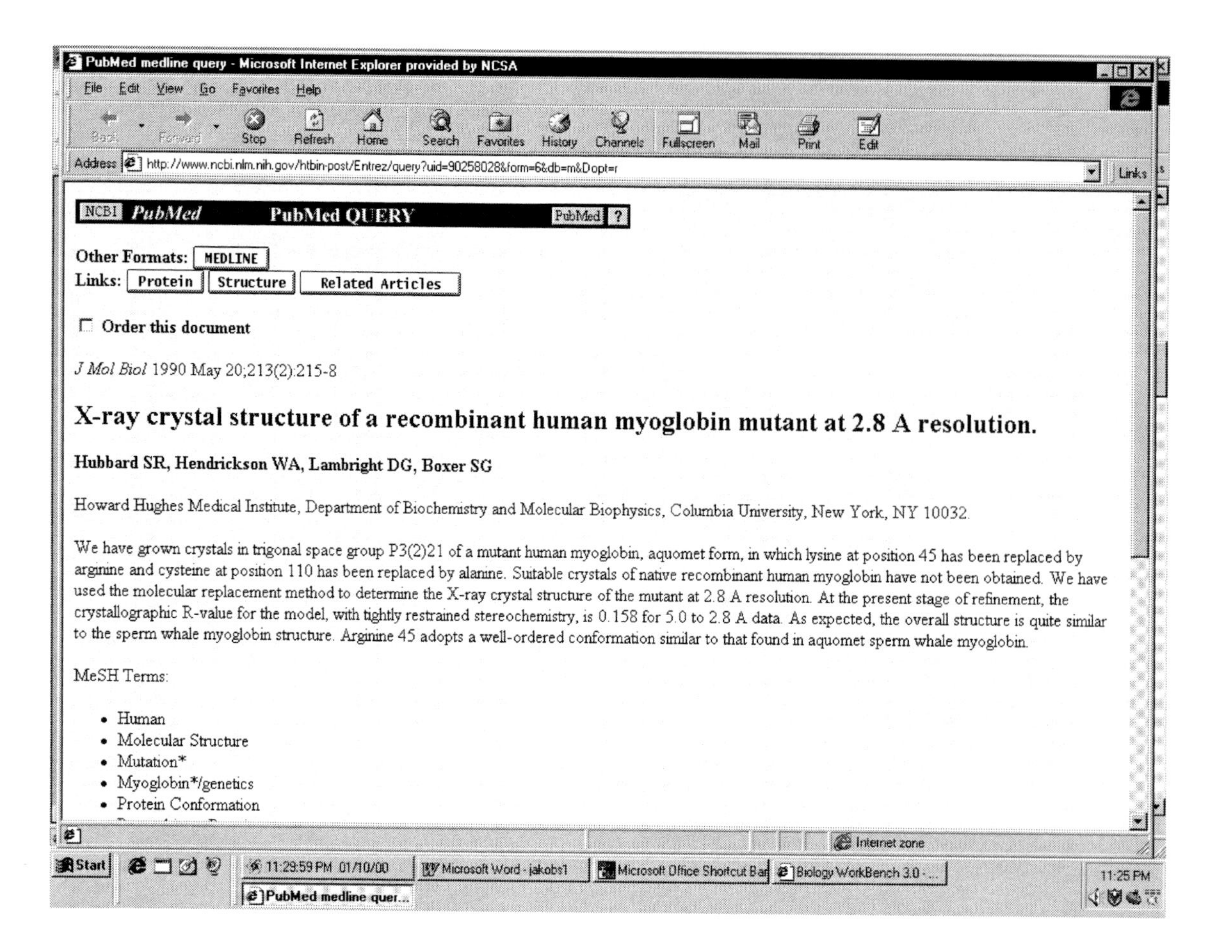

FIGURE 5.7 Original paper for the structure of myoglobin obtained for the hyperlink in the database illustrated in Figure 5.6.

something that you are doing on it. So, people really do need to be guided through using these environments. Of course, this is the great overall problem in the information flood—separating the junk from the good stuff.

People talk a lot about democratization of the Internet. They talk about two tendencies: on the one hand, we should use the Internet to democratize information; on the other hand, there is a big difference in how people use this information—and it depends on economic class, which cuts across the same fault lines as everything else in our society. I think the technology is getting so cheap that the real fault line is not going to be between people who physically have access to it or not; it will be between people who have access to the right guidance in using it or not.

James D. Martin, North Carolina State University: I would like to comment on using this kind of technology in an educational setting. By all means, I use it in the classroom, and I find it quite helpful. But, I also am feeling an increasing distaste for Web evangelists. Maybe it is because I am from the Bible Belt. There seems to be an almost evangelical flair to Web proponents. As one of my colleagues at NCSU suggested, we seem to be creating a click and drool population: we click on it, it looks cool,

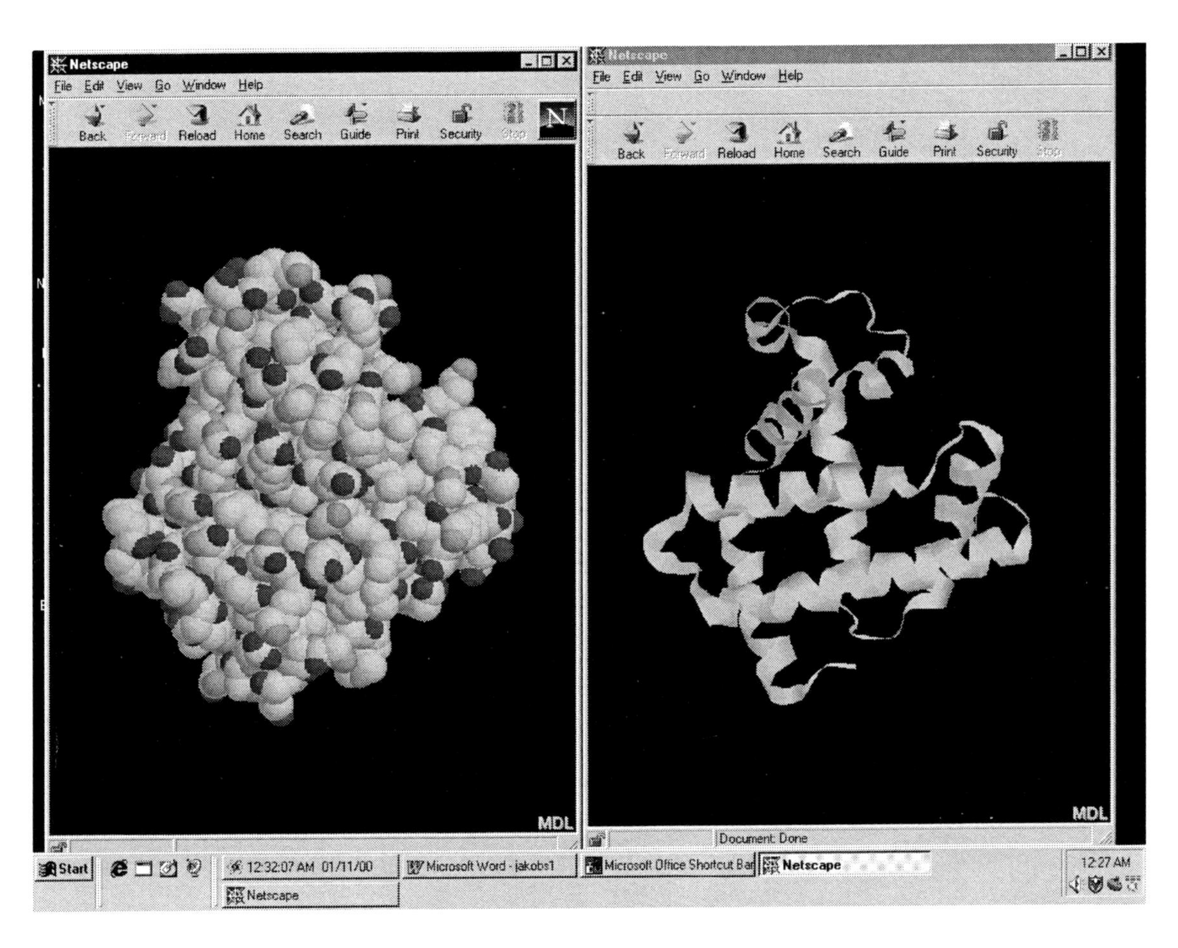

FIGURE 5.8 Illustration of the myoglobin structure. The space-filling structure is illustrated on the left and the ribbon structure on the right. The structures were obtained from a hyperlink from the database seen in Figure 5.6.

and therefore it is good. I also see a lot of research proposals that require an educational component, such as for the Research Corporation or for a NSF CAREER award, for which the educational innovation boils down to putting something else onto the Web. In my teaching of both freshman chemistry and advanced inorganic chemistry at the graduate level, I have two experiences that I think are important to relate.

At NCSU, we have redesigned our freshman chemistry curriculum and have computer support that goes with our lectures that has animations and demonstrations that I can't do on the blackboard. They are extremely useful. But at the end of the semester I had a number of students crying for me to go back to the blackboard, for a simple reason: what I wrote on the blackboard could be copied into their notes. This is an accessibility issue that affects a student's belief that "one day I am actually going to be able to do this myself." By contrast, click and drool demonstrations tend to be so beautiful that we become users only and don't really believe that we could ever do them ourselves.

In my graduate course, advanced inorganic chemistry, structural aspects are extremely important. And thus, I try to teach visualization techniques. I start with point groups, then teach space groups and

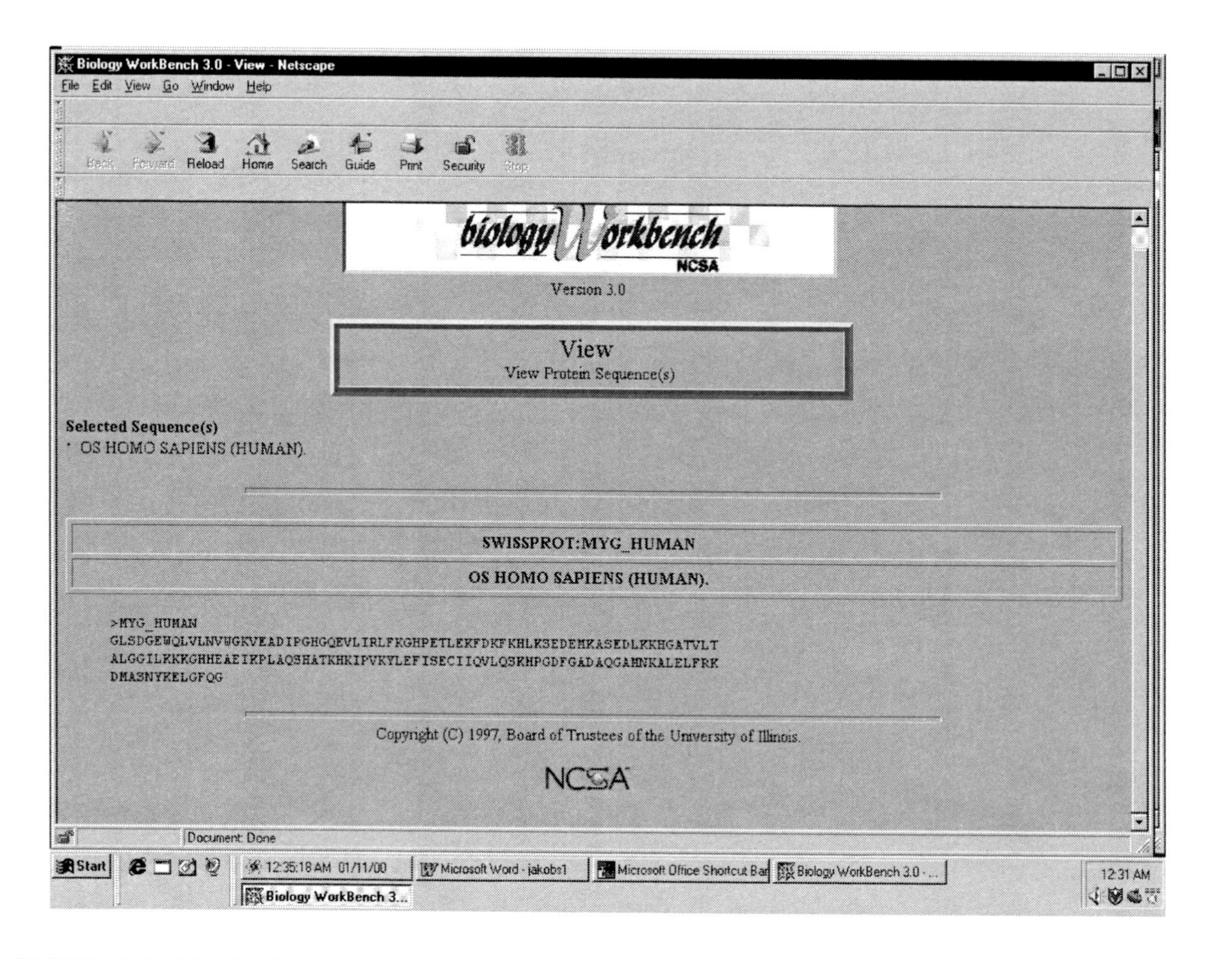

FIGURE 5.9 The database of the sequence of amino acids is also available from the National Computational Science Alliance Biology Workbench, as illustrated here.

the whole works. I try to get the students to learn how to take a crystal structure that is reported in a journal, draw a picture of it, rotate it, and to get a feel for what structure is like. I wasn't succeeding in getting many people to get a real grasp of it. Then I asked myself what it was in my training that gave me this ability to connect 2D to 3D, which we do, whether it is on a computer screen or in a textbook.

What suddenly dawned on me was that my wood shop classes in the seventh, eighth, ninth, and tenth grades were probably the greatest asset to my education in terms of my being able to connect two dimensions to three dimensions, a skill that is absolutely important to chemistry. My seventh-grade shop teacher would not let us touch a tool until we had drawn a set of blocks to learn drafting skills, so that we could perceive two and three dimensions. I said, what the heck, and I got the guys in the shop to create a set of blocks. For advanced inorganic chemistry, students had to draw two perspectives of a set of blocks. I was amazed. They had a much better grasp of dealing with the computer screen after using the blocks. And on exams—I also teach molecular orbitals—I actually got pictures that I could interpret, as opposed to some kind of a scrawl.

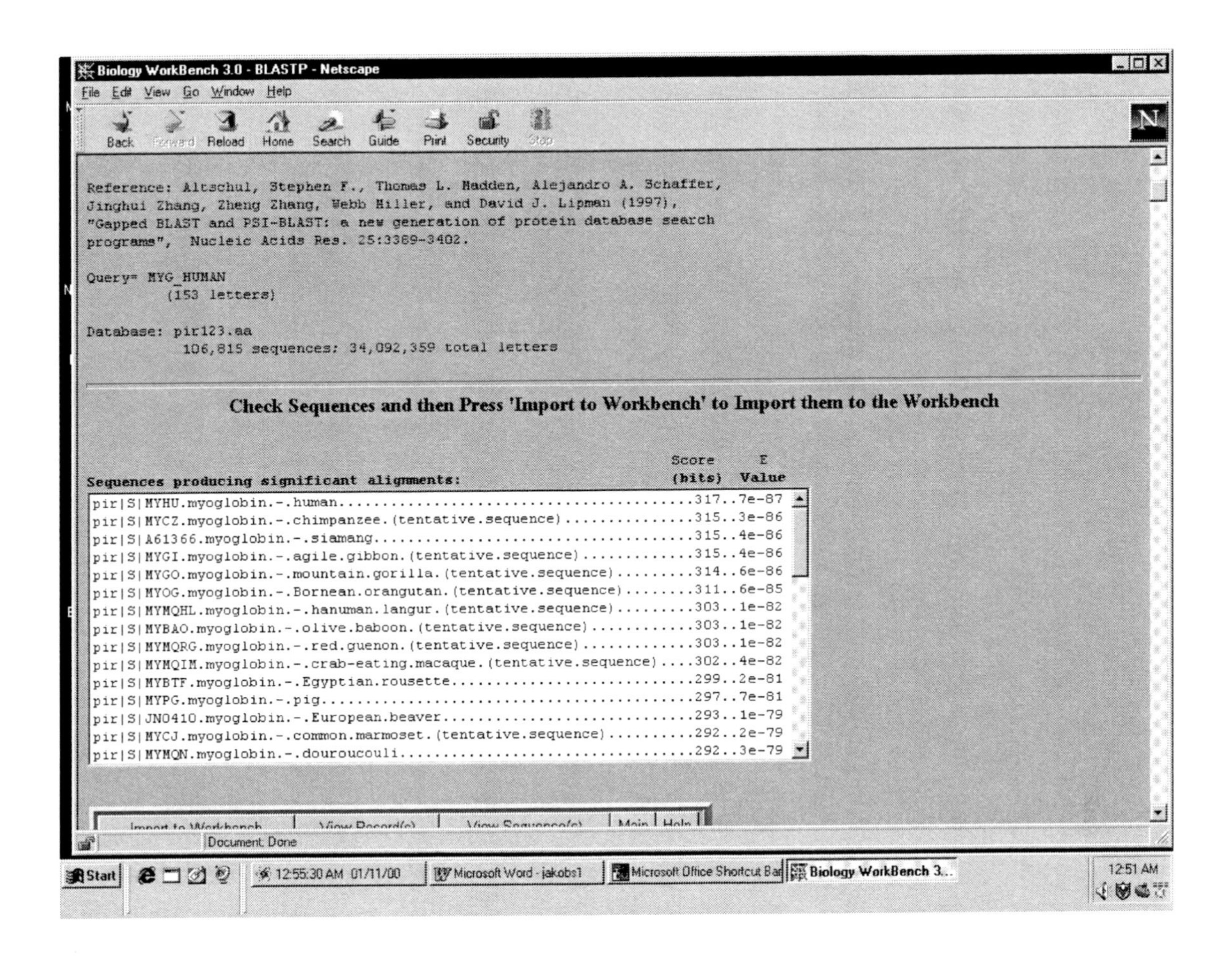

FIGURE 5.10 The results of an operation called BLAST on the sequence in Figure 5.9 are seen here. This operation involves sliding the given sequence across all the thousands of sequences in the database of Protein Information Resource, and finding the ones that are most closely related. This figure shows the "hits"—the most closely related sequences.

I guess I am calling for a bit more moderation. Let's move away from the evangelical Web and recognize where it is and is not useful. But also remember that the "traditional university" is this amazing beast, that by definition will be changing all the time, but that has also lasted in an amazingly constant format for thousands of years. I am not sure that we have to reinvent things quite as much as we sometimes hear. Let's take advantage of what is there, an institution that has weathered time and many changes and is going to stay. I think Web moderation as opposed to evangelism is going to be the key.

James S. Nowick, University of California, Irvine: Let me start by saying that the example you presented was of the highest caliber as a pedagogic exercise for students. One of the problems, as we develop sources of information for students, is similar to that of drinking from a fire hose. As educators, we need to teach them how to sip from the fire hose. This is what I try to teach them in my courses. One thing that can impede the portability of what you have developed is that the instructor must be involved in the creative process to do a really good job. If I simply take someone's curriculum resources and try

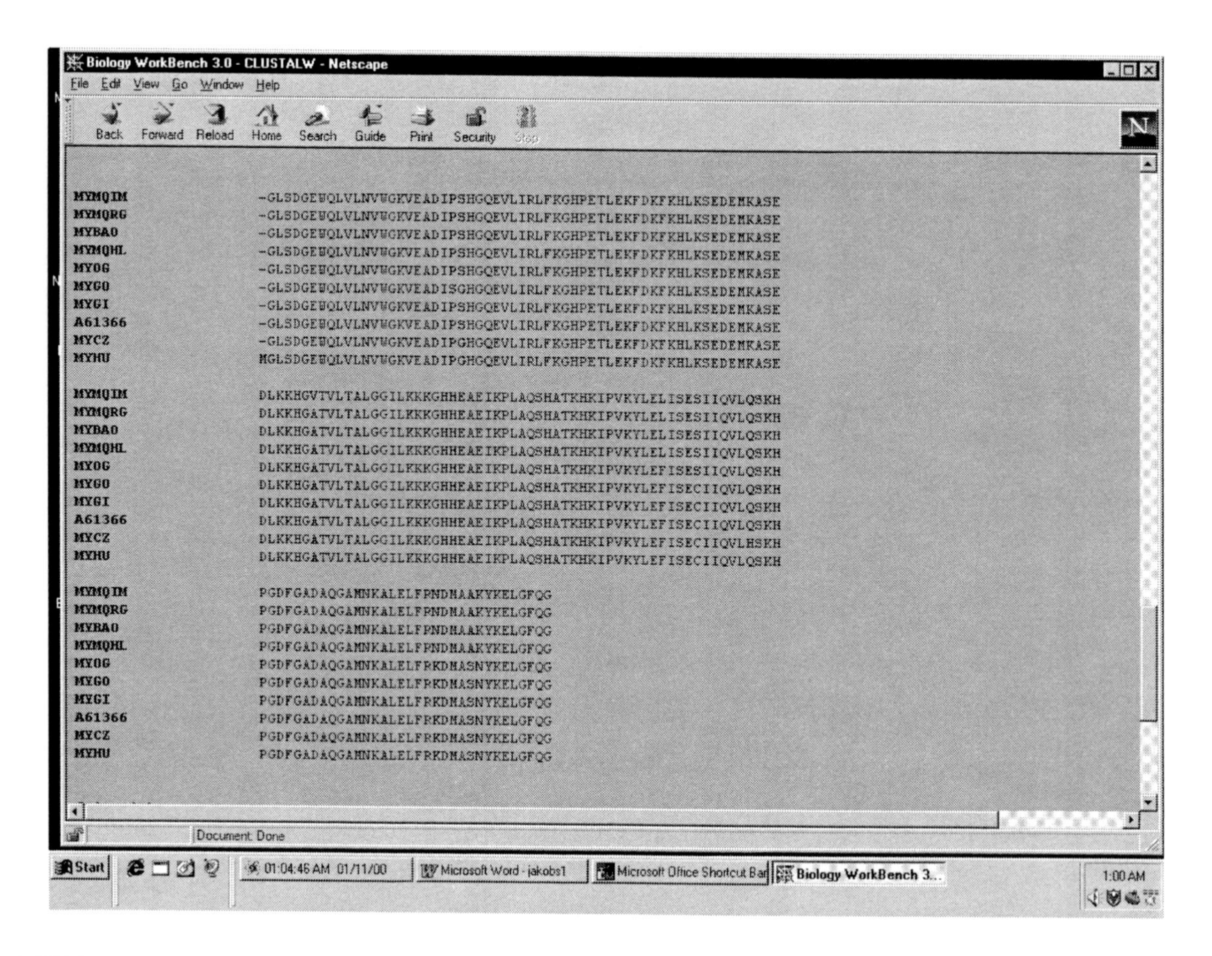

FIGURE 5.11 The sequences of all of the primary hits can be imported into the Biology Workbench workspace and aligned with each other as is illustrated here. The alignment is shaded more lightly when all the sequences are identical with each other and printed darkly when there is variation from one to the other.

to implement them in my classroom, without being personally involved in the creative process, I will do a mediocre job. If I have an intimate role in the creative process, then I will do a great job. As this stuff propagates to your first generation of biology instructors, and then to their peers, part of what will determine the success is how much they embrace this as something that they have put their own brains into.

Eric Jakobsson: Let me comment on that and reinforce what you are saying. That is exactly why we wanted to engage a substantial community of people in developing materials, as opposed to doing that ourselves. I agree with that. I also agree with the call from the previous speaker, to be as interested in what the Web can't do, and what it needs to be supplemented by, as what it can do.

Frankie Wood-Black, Phillips Petroleum: I want to make a comment with regard to the fire hose aspect of the Web and the ability of the Web to do some interesting operations. One is that the Web

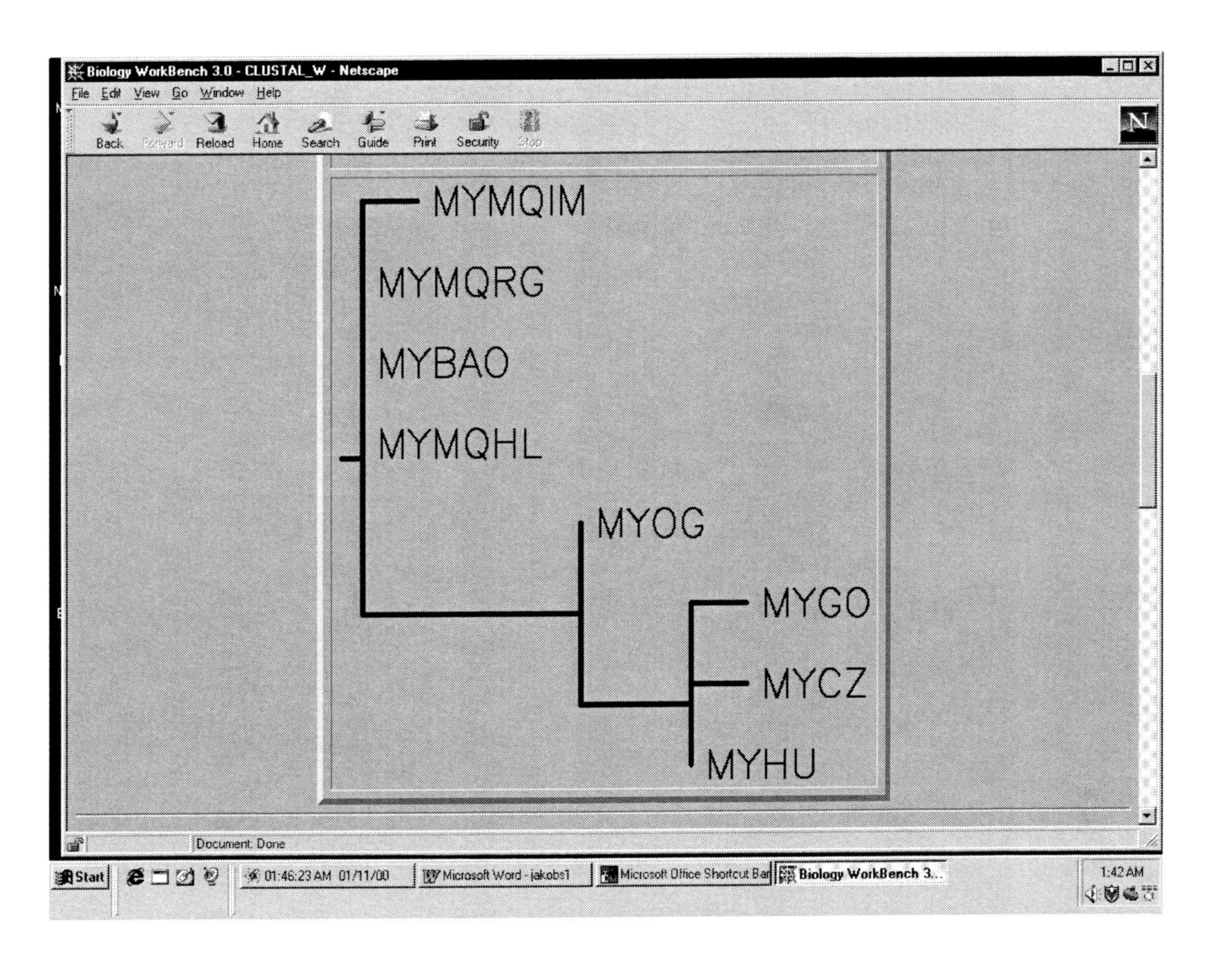

FIGURE 5.12 A phylogenetic tree can then be constructed from the pattern of differences in the alignment above in the Biology Workbench. The total length of the branches between each pair of species is proportional to the evolutionary distance between them.

provides access for everyone to get information. Increasingly, we are passing the point where it is the haves and the have-nots in terms of access but rather having the mentor or guide on how to use it. You have got to build on the fundamentals of where the understanding comes from, and they are doing it now; for example, you have a third-grade class that is sent to the library to figure out what a resource is. The Web is only one resource that can be built upon. The user needs to develop the capabilities of sorting through what is good and what is not good for an overall analysis. We have got to be sure that we are providing the guidance as well as the information.

Eric Jakobsson: I don't intend this to be evangelic, but I think the power of the Web is overwhelming. We talk all the time about the bottom line. The bottom line is that, in terms of how many bytes of information you can transmit per dollar, including dollars for infrastructure, this is a much more powerful technology than, for example, the printed word, in addition to the various things that it can do.

It is like a fire hose. We are all in the stream of this fire hose, and we are not going to turn it off. It

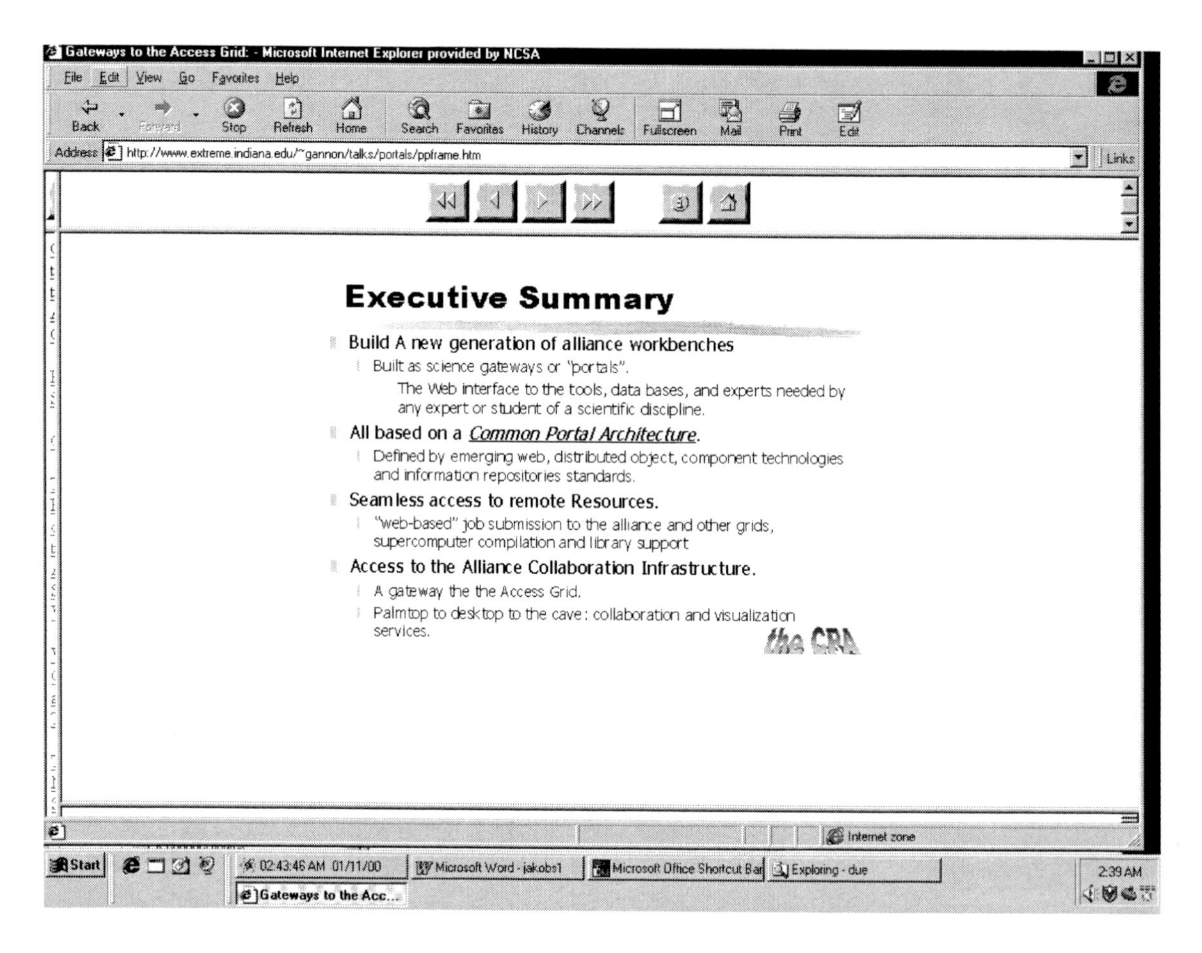

FIGURE 5.13 Executive summary of the National Computational Science Alliance Common Portal Architecture Project, taken from the Web site of Dennis Gannon of Indiana University, the "roadmaster" of the Alliance effort in this area.

is a question of diverting the "stream" to a useful problem—it is an engineering problem analogous to how you are going to get hydroelectric power from a waterfall. You are not going to turn the waterfall off, but you figure out instead some system of turbines and generators, and so forth. We have got an information flow that we have to figure out how to engineer into a useful information technology.

Ernest L. Eliel, University of North Carolina: I find myself somewhat in tune with the previous speakers. I think one of the problems is that we haven't yet managed to map a computer network on the neuronal network in our minds. I think that causes a certain amount of problems, and let me give one specific example. You replaced, in one of your early slides, a disciplinary staircase with a disciplinary framework. My experience is that when I go one step to the right or one step to the left, I have serious vocabulary problems. When I go two steps to the right or two steps to the left, I have very serious conceptual problems. I am not sure that the computer could help me with that. Certainly, I could push the help button, but I am not sure that the right answer would come up.

Eric Jakobsson: Actually, we had a dialog on this, at one point, in *Physics Today*. The question was about what physicists could contribute to biology. My former colleague Gregorio Weber, who has passed away, said a wonderful thing about biology and physics. He said that if you get a physicist with N good ideas together with a biologist with N good ideas, then between them you will get 2N. If you can get all those good ideas and understanding in one brain, then you are going to have N factorial good ideas.

We are beginning to do this, but we increasingly have to train people earlier, at least some of our young scientists, in an interdisciplinary mode, as opposed to expecting physicists and biologists to get together later in their careers, or chemists and physicists and biologists.

Timothy A. Keiderling, University of Illinois at Chicago: I would like to speak from this morning's point of view and note how this presentation might reflect back on it. We talked about the need to speed things up and the need to be multidisciplinary in graduate education. It is the kinds of tools made available by information technology that can allow us to maintain a shorter time frame and let the students take their physical or chemical backgrounds and address the biological problem.

This kind of structure visualization tool allows my students to get to the real science more quickly. They used to dig up a structure on the Protein Data Bank and then run around finding the paper, then run around trying to see if they could find a program that could tell them the secondary structure, and finally, they would try to find some answers to a structural question. Now, if they can push some buttons in a unified software package, they can ask the scientific questions that they are trying to answer, and they can stay on theme better. That will allow them to meet these requirements that we were asking for this morning; that is, to keep on a reasonable time scale and yet still be multidisciplinary. I think these technologies have a powerful role. At the same time, like many of you, I am a bit frightened, since I worry whether students learn anything with this quick process. We must make them learn the basic skills so they can utilize the technology meaningfully.

Eric Jakobsson: You know, all new information technologies are frightening. From my reading—I am not a historian—my understanding is that there was a tremendous issue around the invention of the printing press, and how socially dangerous it would be to let everybody read the Bible for themselves, without the proper vetting by the ecclesiastical authorities. This is exactly the kind of question that has been dealt with before.

Dale Poulter, University of Utah: I happen to come from one of those western states that is very conservative and careful with their money. Our governor is the one who instituted the Western Governors University. For those of you who have not heard of it, this is a way to get an undergraduate degree on the Web. I think there is going to be a push into the graduate area as well at some point. Clearly, it is a way to transmit information, and to correlate information, and there is no denying that. The real concerns I have are about socialization, because although chat rooms are nice, I think humans need the one-on-one visual contact with another person to become properly socialized.

I have the same concern about mentoring. Effective mentoring has to be done up close and personal. The other problem that may not be recognized is that when you deliver a Web-based class, you don't save time. It is almost like a British tutorial in that you can't answer or address a single question to 50 people simultaneously and get instant feedback. You have to do it one by one by e-mail. It takes a tremendous amount of time.

Eric Jakobsson: I second the motion. Getting involved with these types of issues is not a way to work less hard.

John Schwab, National Institute of General Medical Sciences: I have heard people here comparing the Internet-driven information glut to a fire hose. This is a valid metaphor in terms of quantity, but not in terms of quality. The water that comes out of a fire hose is of uniform quality. What comes off the Internet is quite variable; there are good data and lousy data. There are good applications and flawed applications.

Unfortunately, there has long been a tendency, even in pre-Internet days, to think that if data were obtained from a machine, that those data were somehow validated. Clearly, that is naive. The utility of the Internet as a learning tool depends ultimately on how knowledgeably it's being used. This leads me back to Dale Poulter's point regarding the importance of mentoring. It's absolutely critical that Internet users be taught to question and evaluate the quality of the information that they are accessing.

Eric Jakobsson: I would even go further than that. There are not only bad data and bad applications, but there is also evil. There are hate groups. We learned a couple of generations ago that there is nothing incompatible between evil and technical competence. That extends to the Web as well. It is up to people who want to, and are capable of doing good things with it, to take it over and use it as well as possible.

6

The Graduate Student in the Dual Role of Student and Teacher

Angelica M. Stacy
University of California, Berkeley

Beginning graduate students face many challenges as they adjust to a new learning environment in which they will assume the dual role of student and teacher. The goal of this discussion is to raise awareness of the challenges facing new graduate students so that we can think in new ways about how we offer them support and professional development. Specifically, we want to discuss the following questions:

- What models of teaching and learning might graduate students hold?
- What assumptions do we make when we place graduate students in the role of instructor?
- How can we enhance the experience of graduate student instructors?

WHAT MODELS OF TEACHING AND LEARNING MIGHT GRADUATE STUDENTS HOLD?

It is reasonable to assume that new graduate students will draw many of their models of teaching and learning from their prior experiences in school. Thus, we want to consider the way in which chemistry instruction typically is offered. These new graduate students have spent many hours at their undergraduate institutions sitting passively at lectures. While the quality of the faculty presentations might have been quite high, it is likely that little time was available for the students to think about ideas. Many spent most of the time copying directly from the board into their notebooks.

What is the model for teaching and learning that one might draw from observations of these behaviors? Certainly, there is a belief that communication of ideas in chemistry is facile. If we explain these ideas carefully, in the way in which we understand them and have organized them for ourselves, students will understand them. We assume that the vocabulary we use has meaning, that students have mastered the foundation of understanding on which we are building, and that they believe what we say. It all seems so efficient. By simply listening and taking notes, students can process the information and learn new ideas.

We might refer to this model of teaching and learning as the "telling" model. Teaching is about transmission of ideas. To learn you need to receive and memorize these ideas. Another way to say this is that students come into our classes as blank slates on which we write information. What are the learning outcomes? Students repeat ideas and often sound approximately correct when we ask them about what we taught them. But what have they really learned?

In her Ph.D. thesis (University of California, Berkeley, 1999), Melonie Teichert showed that the lecture model was not successful in helping students understand that energy is required to break bonds. She reports that despite the fact that this idea was stated explicitly several times during several lectures, many students believed the opposite. Indeed, Teichert found that *fewer* students were able to answer correctly that energy is required to break bonds on a posttest at the end of the semester compared to a pretest given at the beginning!

It is tempting to try to dismiss these data. Perhaps the students are not very good. Their backgrounds are poor. The explanations were not complete. I assure you that the students involved in Teichert's study are talented, bright, and capable. The faculty instructor is a dedicated teacher who received high ratings from students and from faculty observers. No, it is not possible simply to make excuses about preparedness of the individuals involved. We need to think more deeply about what might be going on, and look beyond trivial assumptions we might make to dismiss these data.

It is interesting to consider why students might draw the opposite conclusion about bond energies from what we tried to teach them. Why might students think energy is released when bonds break? Consider their prior experiences and observations that they have made. Students have watched paper burn. As the paper breaks down, energy is released. The paper is gone, and to the observer, it appears as if nothing is left. Students speak about getting energy from the foods they eat. Again, as the food is broken down through digestion, energy is released. In my opinion, students are making reasonable conclusions on the basis of these experiences. When things break down, energy is released. Why should they believe that energy is required to break a bond just because we tell them? It simply doesn't make sense given these prior observations and little counterevidence.

There is plenty of other evidence in the literature that students hold ideas that are contrary to those we think we have taught them. The videotapes "Private Universe" and "Minds of our Own" (The Annenberg/CPB Math and Science Collection) contain other cases. An example from these tapes regards photosynthesis. From where does the tree get its mass? A common response from students is soil and water. This answer is given even after detailed instruction about photosynthesis. Students write down the equation that carbon dioxide and water are involved but draw little meaning from it. Again, it is not possible simply to dismiss this example as anomalous. Students have good reasons for the ideas they hold. In this case, it is hard to believe that a gas (which does not crush us) can give rise to something as massive as a tree.

Now back to our entering graduate students. They believe in the "tried and true" transmission model of teaching and learning. After all, this method has apparently worked for them. They are the success stories. They have been told repeatedly that they are the best and the brightest (because only the best and the brightest succeed in science). But, indeed, they also hold many ideas that are counter to those they have been taught. They are still struggling to gain an integrated knowledge of chemistry beyond doing well on tests that often emphasize memorization. It is these students with these experiences whom we now place into the classroom to teach undergraduate students.

WHAT ASSUMPTIONS DO WE MAKE WHEN WE PLACE GRADUATE STUDENTS IN THE ROLE OF INSTRUCTOR?

Graduate student instructors have just come from an environment in which faculty talked at them. Now, they have to adjust to a new living environment, figure out the university's bureaucracy, excel in graduate courses, find a research advisor and, by the way, they have to teach.

When we ask new graduate students to go into a classroom and teach, what assumptions are we making? Our actions imply a set of beliefs, which include the following:

- They know how to teach (after all, they have been students).
- They understand the way the class runs (even though they have come from diverse undergraduate institutions).
- They are comfortable in being partially in charge (they have to assign grades but as proscribed).
- They are perfect slates of chemistry knowledge (they have an integrated understanding of chemistry).

Are these assumptions valid? How successful would you be in the following situations:

- You watch a carpenter, then teach carpentry. (Can you learn how to teach only by watching others?)
- You read books, then teach writing. (Is it enough to have read about science without having been engaged in doing real science?)
- You take 4 years of Spanish, then teach Spanish. (How well do you need to know the language before you are effective at teaching it? Aren't students struggling with a lot of new chemistry vocabulary words?)
- What if you need to teach and English is not your native language?
- Could you imitate the teaching of a colleague? (Can you be a clone of any of your colleagues?)

Many of us would not be comfortable in the situations described above, yet we are asking the graduate student instructors to work in situations analogous to these. They have only watched others teach, and then we ask them to teach. They have read about science and done directed laboratory experiments, but have done very little research, yet we expect them to know what science is all about. They are struggling with the chemistry vocabulary they have heard over the last few years, and now we want them to teach others to speak the language of chemistry. Moreover, they have to teach in the way that we dictate, which may or may not be consistent with their own views about teaching.

Despite the high regard in which we might hold entering graduate students, it is clear that they may not have the knowledge and resources necessary to meet the demands we place on them. The realities are that they were recently undergraduate students themselves, and now we are asking them to be the instructors. They are uncertain about protocols in their new environment, yet they are asked to guide undergraduate students. They have to assign grades as proscribed to students they may consider as peers. They have to teach what they are told to teach, even if they are uncomfortable with the material or believe the course should be structured differently.

HOW CAN WE ENHANCE THE EXPERIENCE OF GRADUATE STUDENT INSTRUCTORS?

There is ample evidence that we are placing graduate student instructors in a position for which we do not give them the support they need to be as successful as is desirable. If we want to change this situation, we need to give more thought to the professional development that we provide. To help initiate a discussion of how to do this, I would like to offer two examples of how graduate students have achieved high levels of success in their roles as instructors.

First, let's return to Teichert's study of student understanding of bond energies. The results reported above were for the control group. In Teichert's study, there was a group of students (the intervention group) who received a slightly different treatment. The graduate student instructor for two discussion sections was armed with a worksheet on bond energies that helped guide students to think about their prior conceptions. In the intervention sections, the students struggled with the ideas as they tried to make sense of all the evidence they had. The graduate student instructor facilitated the discussion. In contrast, the graduate student instructors in the control sections did what they were most comfortable doing—they talked at the students about bond energies.

As discussed above for the control sections, student understanding of bond energies declined on the posttest compared with the pretest. However, with the single intervention in which the graduate student instructor facilitated a discussion that engaged the students in evaluating evidence, the results were strikingly different. The students in the intervention sections did significantly better on the posttest compared to how they had done on the pretest.

In another study, Lydia Tien and Dawn Rickey developed a model for instruction in the laboratory called the MORE Frame, which stands for Model-Observe-Reflect-Explain.[1] Instead of performing standard laboratory experiments, the students in the MORE sections spent more time thinking about their ideas and refining them. The students in the MORE sections covered less material and spent less time on calculations compared to the students in the control sections. Despite differences in the laboratory curriculum, all students attended the same lectures and took the same exams, which were based mainly on the material presented in lectures. The students in the MORE sections, who were encouraged to think about their ideas, did significantly better on the standard final exam administered at the end of the course.

These are two examples of the success that graduate student instructors and their students can achieve with sufficient support. If we want to create more opportunities for all of our students to enjoy these kinds of successes, then I believe we need to rethink the telling model for teaching and learning. The successful graduate student instructor described above adopted a different model called a constructivist view. Rather than talk at the students, the instructor took on the role of a guide who helps students construct their own understanding. This graduate student instructor gave students the opportunity to think and rethink and to evaluate whether evidence they gathered was consistent with their ideas. The evidence cited above documents that learning outcomes are much greater with the constructivist model for teaching and learning.

SUMMARY

Beginning graduate students face many challenges as they adjust to their new learning environment. There are difficult graduate-level courses in which they want to succeed, the daunting task of choosing a research advisor, and general uncertainties of a new school and a new place to live. Simultaneous with the major adjustments they need to make as new students, they are given the responsibility of educating undergraduate students barely 3 to 4 years their junior. This responsibility is cast upon them with very little formal training. The expectation is that since they themselves were students, they know how to

[1]L.T. Tien, D. Rickey, A.M. Stacy, "The MORE Thinking Frame: Guiding Students' Thinking in the Laboratory," *Journal of College Science Teaching,* March/April 1999, pp. 318-324.

teach. Indeed, the implicit assumption is that they have mastered the material they are asked to teach, know how to teach it, and are ready for the responsibility of evaluating the performance of students close to their own age.

As we explore the demands upon new graduate students in their role as instructors, we need to consider the validity of the assumptions we make as we place them in this role, and examine the realities of the graduate student teaching experience. We need to consider what measures we might undertake to equip graduate students so that they are better able to meet the challenges they face as graduate student instructors. My view is that this endeavor is pointing us in the direction of rethinking our own views of teaching and learning so that we are better able to serve all of our students.

ACKNOWLEDGMENTS

I would like to acknowledge the many stimulating discussions I have had with Eileen Lewis, Robert Bornick, Joshua Gutwill, and Elaine Seymour regarding the ideas expressed above. The National Science Foundation, the Camille and Henry Dreyfus Foundation, and the Exxon Education Foundation are gratefully acknowledged for support that they have given to address the issues raised.

DISCUSSION

R. Stephen Berry, University of Chicago: I need to say something about your next-to-last transparency. It was a little bit of a contradiction of your own case. I expected you to say, when you were putting that up, that the graduate students, when they start their teaching assistant (TA) training, should to be able to go through the experience of being in the situation of the students they are going to teach.

Angelica Stacy: I absolutely agree. I think that is what I mean. So, here is an interesting thought. It is a very important point. If we want our students to construct an understanding of chemistry, we need to assist teaching assistants and faculty to construct their own understanding of teaching and learning, and there are multiple correct answers, if you will.

Judson L. Haynes III, Procter & Gamble: While I was at Louisiana State University (LSU), I had the opportunity to teach an analytical laboratory chemistry course for a year. One of the main things that I noticed or observed is that, in chemical education, especially with analytical chemistry, most of the students were premed students who don't care about chemistry. You are a TA trying to explain chemistry to someone who only wants to know the answer. They only want to know if they got a grade above 98, because they are going to medical school and need the grade. So, it discourages the one or two chemistry majors who are trying to learn chemistry and are truly interested in it. I found it to be a big problem.

Also, the experience of being a teaching assistant immediately upon arriving in graduate school was overwhelming. I went straight from undergraduate studies to teaching a class with 30 undergraduates. That is an overwhelming feeling. I think one thing that can really help is having a system and orientation. That is one thing that we had at LSU. You come in a week or two before you start school, and you attend an orientation. You get training before you begin. Universities need to take those things into consideration.

Angelica Stacy: You liked the training, in other words.

Judson L. Haynes III: Yes, definitely. If you don't get that training, or you get one day and walk out feeling overwhelmed, then you think it is only you. If you spend more time with other people, then you find out they also feel overwhelmed. Then we can bond together and strengthen whatever needs strengthening.

Anne Duffy, University of California, San Diego: I have had quite a bit of experience as a teaching assistant. When I first started, I was not a typical TA, because I had had training in my workplace with employee orientation. So, I could communicate when I started, which I think is a big hurdle for many TAs, because they have a hard time getting their points across. One thing I wanted to say, though, as far as my learning experience, is that I first taught organic chemistry. I had had a lot of organic chemistry and worked in a lab as well. So, I knew my subject and felt really comfortable, even with premed students.

Then I got thrown into a physical chemistry class and had to teach quantum mechanics. Although I had done well in quantum, I had all of a sudden to teach it and relay whatever I had done to students. My learning experience was that I was able to acknowledge that a couple of students in the class knew more than I did. So, I really relied on them to help me when I was miscommunicating or when something wasn't coming across.

I find that TAs have a hard time, when they are in a position of authority, admitting they do not know something. They also have a hard time asking for help from the students. That is a partnership issue. That is not an authority issue. I think that they are told that they have authority, which they do, but they are also in a partnership with the students they are teaching.

Victor Vandell, Louisiana State University: It is interesting to me that you would talk about problems with teaching and communicating ideas to students. As an undergraduate, I remember sitting in my classes and noting the arrogance of the professors who presented the information. I always got the impression that I should know what they were lecturing on before I even got there. It seemed to me that if I knew it already, I wouldn't be sitting there. I took the stance that one day, when I would be teaching students, I would be an animated teacher who would do whatever I needed to do to get the concepts across.

In the presentation you just gave, because of the way you interacted with the audience and your animation, I picked up on all the concepts that you were talking about. I think that is also something that needs to be instilled into the teaching methods of the students. My question would be, If you train TAs to know their material better, and maybe get ready to address and control a classroom environment, then how do you also teach them to relate to their students and effectively communicate, especially when, on top of everything else, many TAs are foreign students?

Angelica Stacy: I can say only one thing about that. Communication is two sided. You can think you have the most brilliant lecture in the world, but if you haven't checked with what people are hearing, it doesn't work.

Karen E. S. Phillips, Columbia University: I am particularly interested in this subject because I really want to be a teacher on the undergraduate level. Another thing that I want to add is that my friends tend not to be chemists. They tend to be musicians and artists and in professions like that. Usually, when people hear I am doing a Ph.D. in chemistry, they grimace. In other words, they think that this is one of the worst things in the world. People always come to me with horror stories about trying to learn chemistry in high school or as an undergraduate. I am so glad to hear your choice of words because I always tell people that learning chemistry is like learning a foreign language. If someone is standing up in front of you and assuming that you know the vocabulary, when you really don't, then nothing is going to get across to you.

You spoke about the fact that we, as TAs in graduate school, are expected to learn from the people who taught us how we should then present ourselves as teachers. If the people who taught us are just talking at us, and not really putting us in touch with what we need to know to understand the subject, then that is what we are going to pass on to our students. I have had more experience than the average TA, because I had the opportunity to teach while I was still an undergraduate. Therefore, I felt way ahead of the game when I entered graduate school and became a TA. I was also put into a supervisory position to help train the next year's TAs. This seemed like a mere formality, however, because although I had gone through the training programs with the students and had certain opinions about who would be effective in teaching what areas of the chemistry course, no real use was made of my opinion. There was never a real opportunity to say how the TA training program could be improved in future years. I think something needs to be implemented by departments as a whole to deal with issues such as these. I went through an associate's degree at a community college and then a small liberal arts college. The term "TA" was foreign to me until I got to graduate school. I think many universities are doing a disservice to their students by having TAs who are not interested in teaching or who are not properly trained.

William Jenks, Iowa State University: The problems that graduate students face in selecting a major professor are much like those faced by an assistant professor seeking a research grant. Fortunately, I am no longer an assistant professor, but I remember being one well. My question is about foreign nationals as TAs. The academic culture that they come from is often significantly different, and I am not sure I fully understand the differences. I know that their academic culture is different from the one I grew up in. When you are training or retraining TAs, do you take that into account, and what impact does it have on what your are doing to train TAs in your system?

Angelica Stacy: This is a good point. I think you have to take different cultural backgrounds into account. If you are going to instruct someone, you need to know where they already are. The best thing you could do is guide them to become better TAs. We should understand the cultural backgrounds of the foreign TAs and what their academic life has been like.

Joseph Francisco, Purdue University: At Purdue, in the fall semester, I have to coordinate the training of more than 2,500 students going through freshman chemistry at Purdue and coordinate with five prima donna professors. I want to tell you of an experiment that was done out of frustration about what to do to help everybody, particularly the students and the TAs. Because we have a large number of students, we can get good statistics, and also, if it works, we can make a convincing case to our colleagues as something to consider trying. One of the things we realized, at least I realized, is that one of the problems with my TAs was that although I sat down with them every week to talk about what was going on in their recitation and what I would like to see happen, it never happened. When I made visits to the classroom to see the things that I wanted them to do, they were not doing them, simply because they had a model of what a teacher should do and that was a model of me. I did not want them to go into the classroom being a model of me or any other instructor. That was not the point.

In response, we developed what we called continuous TA training, because we learned from an experiment that, although there are TA orientations or boot camps, TAs don't remember any of it when they get into the classroom. So, we decided we would work with the TAs every week on their teaching.

We went into their classrooms, observed them, and gave them feedback after they came out of the classrooms. We taught them how to empower themselves and take more control of the classroom, not through lecturing the students but by empowering the students. As coordinator of this training, I am able to divide up the TA group any way I want to. So I split it in half—half went through the continuous training and half didn't. One of the things that we found was that the foreign TAs who underwent the continuous TA training were getting very good evaluations. It became clear by the higher grades of the students who participated in continuous training that it was very effective.

We also found that you don't need to do it the whole semester. Continuously training and intervening up until about the middle of the semester is all that they need to have the confidence that they can control the classroom, run that classroom, and instill that thinking. In case people are interested, we published an article about this in the *Journal of Chemical Education*.[2] It came out in the January or February issue this year, I think. I will also say that the faculty became overwhelmingly positive in embracing this, and we adopted continuous training for all new TAs.

Let me make a comment on another thing that you said in terms of the issue of student learning, how students think about things, and how they structure their learning. The key emphasis of your work sheet is on how to get the students to assess their own understanding. There are other tools out there as well to help students do that. One of the tools that we are experimenting with at Purdue is concept learning. What we are finding is very intriguing and, again along the same lines of what you are finding, very helpful to the students. The problem that we are seeing is that textbooks present chemistry in a linear fashion. We all know that chemistry is not discovered linearly, and the concepts are not presented linearly, but this is how the students construct their framework of learning. What you have to do is to break students out of that mold of linear learning and to see where their lack of understanding is. Once you are able to do that successfully, that is when you can have active learning.

Larry Anderson, Ohio State University: Joe Francisco convinced me that I should step up here and say something about the Early Start program that Ohio State runs each summer prior to the first year of graduate school. We offer anybody who accepts our program the opportunity to come in the summer. About 60 percent do, so we have roughly 40 students. We test them when they first come in, and then we give them a set of courses. This all began in the 1950s with Mel Newman, who decided that students needed to know more about synthesis. So, we had this total immersion synthesis course taught by the organic chemists, but also a number of other courses.

We require all international students to come, and we teach them English as a second language. One of the consequences is that the foreign students are better skilled at presenting themselves in their TA duties than our domestic students are. They come in and say, "My name is Win Wa Chang. You can call me Windy. As you can tell from my accent, I am not a native speaker of English, but . . . ," and so on. This has been very effective. We also have a full quarter course, the summer quarter, in teaching college chemistry.

James S. Nowick, University of California, Irvine: How do you get your colleagues to buy into your constructivist model of teaching at Berkeley?

Angelica Stacy: The best we can do is to encourage the research community to stop making assumptions about themselves as teachers and about their students, and to start applying the same strategies that

[2]S. C. Nurrenbern, J. A. Mickiewicz, and J. S. Francisco, *Journal of Chemical Education*, 76, pp. 114-119, 1999.

they use in their research labs to their teaching. Find out what is going on. The more data we can collect and the more ideas we can sprout, the better off we will be.

Peter K. Dorhout, Colorado State University: Angie, I really like the idea of the TA training handbook. The American Chemical Society (ACS) puts together a very nice three-ring binder affectionately known as the "green monster" for local section leadership. In a weekend's time, you are taught basically how to lead a section of 2,000 or so ACS members, at least in Colorado. The bottom line is that the ACS has done a lot to teach its membership how to be leaders. A lot of the information in the green monster can cross over into this particular manual, and I applaud it. After all, we didn't learn how to be faculty members other than by being TAs.

Angelica Stacy: We have also learned a lot from the Lawrence Hall of Science, which does teacher training, about how you can train teachers in an inquiry-based way. We have a lot of great activities that TAs can do to learn about learning and teaching.

Peter K. Dorhout: It is a trickle-up theory, because these individuals become faculty, or they become research scientists in industry, or lawyers, or lobbyists, and all these things are really key. The only thing I would add to your TA training manual is a section on ethics. I didn't hear you mention that.

Angelica Stacy: Yes, that is part of the activities that would be in there.

Robert E. Continetti, University of California, San Diego: As a former TA of Professor Stacy, in the absence of other extensive training at Berkeley in the early 1980s, having a motivated lecturer like her is what it took to inspire me to do a good job as a teaching assistant. Now that I am a faculty member at University of California, San Diego, I am aware that we have a course on teaching chemistry, perhaps along the lines of that described by Joe Francisco, that TAs take during their first quarter of teaching. I must admit, though, that I am not familiar with the topics covered in that course, given by my colleague Barbara Sawrey. Perhaps as a new faculty member I also should have taken this course.

Angelica Stacy: There is a lot of great research in cognitive science. Go watch your elementary school teachers. There is a lot you can learn. It is very exciting, and we shouldn't separate ourselves from it so much.

John Hutchinson, Rice University: Dr. Stacy, as you know, we have been using the sort of constructivist approach that you describe here in teaching chemistry at Rice for quite some time. It is extremely effective. The concern I want to raise is that I think it is a reasonably accurate perception that teaching in an active-learning classroom, particularly with a constructivist view, is more challenging than simply standing up, presenting the material, or solving homework problems. The question is, How do we persuade graduate TAs, most of whom are conscripts in their teaching assignments, that they should work even harder, particularly—to echo what was just said—when they see most of the role models on the faculty being unwilling to make that same adaptation?

Angelica Stacy: These are all good points, and I think we have to work at it. My TA training handbook was to help them have some varied experiences, which would help them construct a new model for teaching and learning. I am not going to talk at them about teaching and learning. I am going to give them some experiences, so they see the differences. Developing those is not easy. Also, understanding what the students know and don't know is a very important part of this.

Barbara Sawrey, University of California, San Diego: Very quickly, Peter, to address your comment, the ACS, through the Division of Chemical Education, publishes a TA handbook, *Handbook for Teaching Assistants*.[3] There also is something I think is even better, and I say that in spite of co-editing the one for the division of chemical education. It is the National Research Council's *Science Teaching Reconsidered: A Handbook*,[4] which I think is excellent. I highly suggest that, if you don't have a copy, you get one.

There are two things I want to say. One is that I teach, and have taught for 10 years, a course for incoming graduate students on how to teach chemistry. We don't front-load their learning of this because they can't recall it in the heat of the moment when they are standing in front of a class at a later point in time. It is much better addressed on a weekly basis. Again, they don't need it the whole term. They need it just part of the term, and it works out very nicely. I mostly get postdocs attending as volunteers. The TAs are conscripts, but the postdocs are desperate for teaching instruction. That means that we haven't been providing this guidance at the undergraduate or the graduate level, or the postdocs wouldn't be there.

The other comment is that, almost without exception, the incoming graduate students come to us excited about teaching. The only two things that I have found that dissuade them from being excited about teaching are a bad experience or a faculty member who tells them it is awful. My vaccination for the second problem is to help them avoid a bad experience.

Angelica Stacy: I want to echo Barbara's comment. There is a lot of good information out there for people to find. Barbara mentioned some of it.

Christopher F. Bauer, University of New Hampshire: We have heard a lot of opinions today on student learning, and it is good to see some data regarding this issue. One of the issues mentioned was the need to give faculty or TAs a compelling reason to take a look at their teaching. I think some of the information you have shown supports this statement. Your data are quite convincing, but we need not look just at your work. There are a number of places within the literature to find numerous examples of the same sort of thing. This raises the question about what we know about what our students understand after they leave us.

Angelica Stacy: Can I please acknowledge that there are some of my colleagues in chemical education in the audience? Please tap into their knowledge base. There has been a lot of nice research that has been often ignored, I think.

Janet Robinson, University of Kansas: What I am going to say will corroborate what some people have already said and then address some of the other questions. One question was, How do you get faculty involved? I think one way to get faculty involved is to pick a model that they are familiar with and show how it might apply to TA training, for instance. Since I am the chemical educator in our

---

[3]K. Emerson, G. Essenmacher, B. Sawrey, *Handbook for Teaching Assistants* (Washington, D.C.: American Chemical Society, 1996).

[4]National Research Council, *Science Teaching Reconsidered: A Handbook* (Washington, D.C.: National Academy Press, 1997).

department, and the only one who is specifically designated as that, it is very clear that I can't do any change by myself. I have to mobilize everyone. This is something that I think about constantly. One of the ways that the model is so effective, as we have been seeing today, is that it includes the apprentice. The apprentice doesn't happen in teaching. You have seen that. We throw them in and close the door. Could we make an apprenticeship model? I think the Purdue model, the one that Barbara just mentioned, and our one-hour credit course for half a semester that I just discussed are all examples of this type of model.

There is no point in talking at the TAs when they don't even know what they are going to be immersed in. There is no way that they can connect what is happening. What we have been doing is talking with them a bit at the beginning, giving them some readings that they can use, and then having weekly meetings. The other key is observing them at work, making notes on pointers, and then every week discussing things that come up. That starts to approach what we do in the research lab, how people get feedback on what they do. Teaching is the loneliest profession, as we all know. No one is in there to see what we do, except our students. Because of that, we don't get effective feedback. I think that has been the strongest thing about our approach—the idea of taking this apprenticeship model and applying it to teaching. I think that is something other faculty can understand. I am still, however, attempting to prove at my institution that this experimental course is worth continuing.

Robert L. Lichter, The Camille & Henry Dreyfus Foundation: First, I hope we can avoid one of several false dichotomies, that of the chemical educator vs. the chemical researcher. We are all chemical educators, but we may do things in different ways or with different emphases.

I'm up here because one of Angy's comments pushed a button. You and others have talked about not blaming students and not blaming teaching assistants. To that, I want to urge that we not blame the faculty. Of course, faculty members should know about research in cognitive science. Of course, faculty members should know that information about and for teaching assistants is available. There is a lot that faculty members should know about. I haven't met one person who isn't dead serious about wanting to do a better job in these arenas. But our structure doesn't make it easy for them to do so, especially when they are extraordinarily busy doing the very things we tell them they are supposed to do, which is to advance the frontiers of knowledge. So I would urge people in decision-making positions to find ways to enhance the capability of these very busy folk to engage with larger educational issues.

For example, it's undoubtedly not too far off the mark to suggest that the majority of faculty members doing chemistry, who barely have time to read the multitude of research journals in their areas, certainly have little time to read the *Journal of College Science Teaching*, or even, I suspect, the *Journal of Chemical Education*, not to mention journals in cognitive science. Perhaps the research journals could have summaries, abstracts, or even just titles of papers and other presentations that discuss these issues, so that faculty members are at least easily alerted to these sources. The *Journal of the American Chemical Society* is exactly that, not the *Journal of American Chemical Research.*

Angelica Stacy: I hope the journal editors heard that. I want to make a comment about not blaming the faculty, because I don't want to do that either. I also want to stop blaming the high schools. I think a lot of us who do that haven't been in many high school classrooms. I think we should know more about it and not just think about what we should be doing better and pointing the finger elsewhere.

Eric Jakobsson, University of Illinois at Urbana-Champaign: I will inject a biological truism, and this is sometimes used with respect to, for example, the ecological effect of pesticides or antibiotics. If you put pesticides into the environment, you select for pesticide-resistant insects. If you put antibiotics into the environment, you select for antibiotic-resistant microorganisms. This also applies to social structures. We get exactly the behavior from our faculty that we select for. I am reinforcing the idea that we have to change. However, if we want to modify the behavior, we have to modify the reward system.

Derrick Tabor, National Institute of General Medical Sciences: I want to speak from my experiences as a faculty member; I am on leave from Johnson C. Smith University in Charlotte, North Carolina. The dichotomy between the researcher and educator is not false. The fact that the reward systems are different speak to that. When the reward systems for both are equivalent, then we truly won't have this dichotomy.

The other thing I want to say about students is that I have found in my teaching that I could tell when learning was going on because they were actively displaying emotions, a whole range of emotions that I experience as a scientist. We don't talk about the frustrations that we, as experienced scientists, have. That is because we have learned to ride out the low points and not ride too high on the high ones. Students are different. They wear their emotions on their sleeves, and they get frustrated. We typically don't acknowledge that frustration, and we typically don't help them. So, I started to understand that, when my students were frustrated, they were learning. They were in cognitive dissonance; they were looking at the old model versus the new model. They couldn't make sense out of this. My job was to help them manage the frustration and help them get through it. It will become clear. I don't know when. Do you want me to just tell you the answer? No, no, I will figure it out. That is when learning was happening.

I think we have to acknowledge the emotional aspects of our profession. We enjoy it. We enjoy it because we know we are going to ride through those low points. We are going to be there sometimes, but we are going to figure this out. This speaks to the power that Billy Joe Evans was talking about. That is where the power comes from, because we know we don't see a thing right now, but just wait, we will do some more experiments and we will get through it. That is what her students experienced when they went out and did their internships. It is the same thing, and we don't prepare students for that. That, I think, is one of the real joys.

Victor Vandell, Louisiana State University: I would like to reemphasize a point made earlier relative to the training of teaching assistants. As the young lady from Columbia stated, the current TA training process is a disservice to graduate students. I would like to point out that there is a level of frustration that is apparent in watching a teaching assistant. We are thrown into a shark pool. And that frustration carries over into the teaching process. Some TAs even feel—as I have heard them say when they first come into the program—that they were duped. No one told them they were going to have to teach a class without preparation. It seems to me that the overall process is a type of hazing. There is a mentality that I had to go through it, so now you have to go through it. We need to break that chain. We need to say to ourselves, "Let's just stop right now and rethink what we are doing." Then we have to stop talking about it and make it happen.

Bettina Woodford, University of Washington: I have two brief comments. The kind of experimental course that Dr. Stacy was talking about, and so many other initiatives that were discussed today, seem to be pockets of innovation and change in doctoral education, that we would love to have written up for our Web site (Pew Charitable Trust, "Re-envisioning the Ph.D." project). That is the only plug I am going to give for that, and I am not evangelizing. I want to remind everyone of an observation I made regarding a study in which I was involved. The study, funded by Pew, followed doctoral students for 4 years across eight different disciplines at three universities, two or three being in the natural sciences. What we found was that, in the natural sciences, characteristically, TA training tends to drop off after the second year. Apparently, most of the TA training that people do is in the first and second years of their TA-ships, largely assisting in labs, and sometimes, if they are lucky, getting a lectureship of some sort.

By the fourth year of our study, we were interviewing the graduate students as they were nearing the point of entering the marketplace. They were beginning to understand the realities of the job market and that some of them were going to have to consider jobs where they would be doing considerably more teaching compared with the amount of teaching their advisors were doing in the research universities. To their dismay, they also realized that they had spent the last 3 or 4 years of their training not practicing and preparing the skills necessary to teach at that level. The model of graduate education and, in some respects, the way it is funded, unintentionally preclude the graduate students from receiving training in teaching. This is not helping them, as they will head off in a year or two and be expected to teach three or four classes. That was something I wanted to remind everybody of and to offer as an agenda item if we are considering changing doctoral education.

Michael Doyle, Research Corporation and the University of Arizona: Let me offer an observation. For the last five years, the Research Corporation has operated a teaching and research award program called Cottrell Scholars. We expect a teaching proposal and a research proposal from those who are applying to that program. The research proposal is filled with elegant ideas and complemented by references that give priority to the literature and to those who are practicing in the field. The teaching proposal rarely gives a reference, gives ideas that are generally personal preferences, and says nothing about continuity. I wonder why.

Abraham Lenhoff, University of Delaware: I guess all the faculty in the room agree that we wouldn't have this problem if the TAs emulated us perfectly, because we set a perfect example. One of the paradoxes is that we are spending time talking about fostering teaching, whereas much of our time is spent mentoring students in research. A parallel model for that would be graduate research assistantships for mentoring undergraduates in the research lab. This gets undergraduates going into research, and it also has several benefits for the graduate student. The first and most obvious is that they can get the undergraduates to do some of their research for them. Second, and more important, they are teaching material that they presumably know a lot better. They don't have to go out and learn the material. Third, and most important, is that there is better two-way communication in this one-on-one teaching. The mentor gets faster feedback on how well he or she is teaching, and this leads to an improvement in teaching styles, I think, which can be carried over to the classroom. Perhaps, in addition to the TAs, research assistants, and outreach assistants, we can have mentoring assistantships. The graduate students in my lab who have had undergraduates work with them really have appreciated that opportunity.

David Oxtoby, University of Chicago: I would like to comment on Lynn Jelinski's talk and then describe a program at Chicago. One of the things described was the relationships with industry and the outside world, which mostly tended to be one student going off and doing a project. At Chicago, we have developed a team-based approach. It is a team of students from the physical sciences from several different fields, such as chemistry, physics, even computer science. The other member of the team—and this is important—are business school students. We work with a company on a project that has both a technical and a business side. This has been very exciting, because the team functions as a whole. It is not that the science students are doing the science and the business students are doing the business. Each one is looking at both sides of things and learning quite a bit. They also have that team experience that is very important as well. The teams have a coach who is not a faculty member, but a professional who knows something about science and business. It has been an interesting project, to which the students have responded very well.

There are issues, though, specifically in science cases, of funding this project. The faculty feel that if their students are spending a significant amount of time participating in this project, that they can't support them on their research grants. Therefore, we have to ask the company for extra money to pay the students in the sciences. Of course, the business students are used to paying for their education, but to support the science students, we need to ask for extra money. It is a little bit awkward and not always easy to finance. The university also contributes to the funding. I want to mention that as an example.

Ernest L. Eliel, University of North Carolina: I would like to bring up a new topic, which is not on the program, and that is the master's degree. The master's degree is, unfortunately, now used as a good-bye present in all, or almost all, of the institutions that offer Ph.D.s in chemistry. I think that is unfortunate, because there are many advantages to the master's degree. The first and obvious one is that there is a demand in industry for the master's, in some instances greater than for the Ph.D. One of my colleagues, who works in the area of synthesis, tells me that if any of his students want to leave with master's degrees, they can find a job immediately, because there is a great market for them. In the second place, the master's degree offers the opportunity to take a few more courses. If students take a master's degree, certainly there would be some increase in the number of courses, thus reducing overspecialization.

The third point is that, in my rather long career, I have found that some people get passed through to the Ph.D. degree who really shouldn't get one. The problem is very simple. A student fails either in the first year, when he or she doesn't pass the course exams or, in very rare occasions, fails at the time of the oral candidacy examination, which usually comes at the end of the second year. Even in some instances where I, as research advisor, have urged caution, some of my colleagues have said, "Oh, no, you can't do that." The student has got so far, so let him or her continue on and get a Ph.D. If we make the master's degree an obligatory degree for every graduate student—I would like to throw this in as an interesting idea even though perhaps some of you will not readily buy it—then I think we could avoid this problem.

One additional advantage is that, somewhere in their career as graduate students, students would discuss with their preceptor what they want to do in life. I think that occasion does not always come about, or it comes about late in the student's career when he or she is ready to look for a job. After completion of a master's degree would be the right time to ask whether one should go on to get a Ph.D. Do you really have what it takes? Do you have the imagination that it takes? Do you have the self-starting ability? If not, you might be a good worker in the laboratory and in other ways very clever, but you might be better off ending with a master's degree. I have trained students with master's degrees who have done well in their careers and have had happy lives. I have also had one or two Ph.D.s who haven't been happy at all, because they were advanced too far for what they really had to offer.

Someone said earlier that if you want to become a lawyer, it is a great shame that you spend time and money to get a Ph.D. in chemistry. I would agree with that. I acted as the consultant for a lawyer in a large, multimillion-dollar case. I very much appreciated that this lawyer had a master's degree in chemistry, because we would communicate extremely well. With other lawyers that I deal with, we cannot communicate as well. What applies to lawyers, applies to members of Congress and to many other professionals, who, if they had a master's degree in chemistry, would be very well off.

The objection to that is two-fold. One is that it would take more time, and "isn't it ridiculous, if you want to go on to get a Ph.D., to spend time on the master's degree?" Most universities have two types of master's degrees, a thesis and a nonthesis degree. I think if it is decided that, toward the end of the second year, the student will get a Ph.D., then a nonthesis master's is appropriate. The student then has to write a report, and going over the report in great detail would improve his or her writing and organizational skills, which is something that industry is very much interested in. On the other hand, if the student and the preceptor decide that the master's degree is appropriate, then I think there should be a research thesis, so as to finish the degree in two-and-a-half years and have something to show for it. I think this should be considered very seriously. It would be in the interest of the students, and it might be in the interest of the community as a whole.

Now, there is another rather serious drawback that I cannot conceal, and that is that universities usually pay, out of departmental funds (usually TA funds), stipends for students in their first and sometimes second year. Then, later, when they get into research, they often are supported by research assistantships. The universities would certainly say they are wasting the TA money in the first 2 years if the student doesn't stay on to do a full research program. I think that possibility will have to be faced.

Joel I. Schulman, Procter & Gamble: Before I say what I was going to say, let me echo what Ernie just mentioned regarding master's degrees. Certainly, in industry, master's degrees can be highly valued, particularly if they are thesis master's degrees. I don't think a master's degree that involves only course work is particularly useful to industry. Research experience, even if it is relatively short, can be very useful.

What I want to say, though, goes back to what a previous panel member, Dr. Jakobsson, said about information technology potentially leading to a generation of people who are "multiple experts." That bothered me when he said it, and I finally figured out why. I react negatively to the concept of training a Ph.D. chemist to be a "multiple expert." Rather, with 60 or so percent of Ph.D. chemists going into industry, one of the emphases that should be placed on graduate training is that Ph.D. chemists need to know how to work in multidisciplinary teams. We are not saying that people have to be multidisciplinary themselves or "multiple experts." When you work on a multidisciplinary team, what we are asking is that an analytical chemist be able to work with a medicinal chemist or a physical chemist or a pharmacologist or somebody in business.

Multidisciplinary is different from "multiple expert," and I am not sure that it serves the Ph.D. well to train "multiple experts." Remember the team model, for which you want people able to act broadly and to show that they have depth in a particular area. The idea is that they can drill deeply in another area if they have to, but they can also work broadly with a diverse group of people.

Eric Jakobsson: When I look at the knowledge that I use today, the fraction of it that I learned in school is pretty small. Lots of what I use today is knowledge that didn't exist when I was studying at a university. One of the problems that I had, when I was a young engineer, was coming to grips with the massive layoff that I saw of people who were more senior than I. I realized that this was something that occasionally happens in industry. Because of its bottom line, to a certain extent industry treats scientists and engineers as disposable. That is, if somebody hasn't kept up, if somebody's skills have become obsolete, if the skills of new graduates are more relevant, then the rational thing for DuPont or Air Products or any corporation to do is to lay off the mature scientists and hire new ones. I think the ability to acquire multiple and new expertise throughout life, throughout the entire working life, is really the only safeguard, or the best safeguard, that a scientist or engineer has.

Joel I. Schulman: I agree with that. What I am saying is that, in graduate school, you don't need to be a multiple expert.

Eric Jakobsson: You had better learn a mind-set that will let you do that.

Stanley I. Sandler, University of Delaware: First, I want to thank Angelica for her marvelous presentation, which showed a lot of enthusiasm. I must admit, however, based on my experience in educational pedagogy, that many of the methods proposed do not transfer well. I believe the reason is that students generally are responding more to the enthusiasm of the people involved rather than to the specific methodology used. This suggests to me that the enthusiasm of the instructors is important, more than the methodology.

I also have a question for Eric. I am concerned about intellectual property issues. In your presentation, you showed links to many different sites including databases that have subscription/membership fees or commercial software. It is unclear to me exactly how someone who didn't have access to these sites would be affected, or whether such users were gaining access to such sites by using your site.

The third comment is to Ernest. When I was department chair, I set up an industrial advisory committee, which included vice presidents of various corporations. The question to them was, Should the department offer a master's degree, since it was not a cost-effective research activity for the department? They were unanimous in saying we should continue to offer the degree. However, before we agreed to that, they were to contact their human resources people, have the records of their employees studied, and we would discuss the issue again at our next yearly meeting.

A year later when the advisory board met, they were unanimous in the view that there was no discernible value to the master's degree. In fact, starting in industry directly out of college and getting a year or two of extra experience, instead of spending time getting a master's degree, was more important for long-term career success. So, unlike what has been said here, it is questionable as to whether a terminal master's degree is useful, at least in chemical engineering.

Eric Jakobsson: Let me respond briefly to the intellectual property issue, which is a serious issue, and one that we spent more time on than we had hoped. Everybody we access for databases, or whose software we use, has agreed to let that be used in a public domain way for nonprofits. So, we make the Workbench freely available to everybody who is in a nonprofit institution, either a governmental or educational institution. On the other hand, the Workbench is licensed to corporations, to be mirrored behind their firewalls. They often have to make some sort of arrangement with the people whose data are incorporated. Intellectual property with respect to software and information technology is one of the areas that is in the Wild West of the law. We patented the Workbench. I think, and I believe Shankar does as well, that whatever else we do will be public domain, an open source, and we will never patent anything again.

Timothy A. Keiderling, University of Illinois at Chicago: Angelica said something that stirred me up, but I think it applies to a lot of things we heard today. It was her concept of blame. Her phrasing was very gracious: "Don't blame the high schools." I honestly think what she did was to blame the faculty. She seemed to blame the faculty because she in effect said, "They are not doing what I do, and my way is new and exciting." I think we have to get into those faculty's heads, too. As a department head, I have had to sit down and try to find solutions with people who have taught freshman chemistry and physical chemistry for 30 years. They really know the material and have taught it effectively for a long time. They tell me that they sometimes use the same exams from 30 years ago as a test of the continuity of the course and find significantly lower scores with the current students. The point is that the students are different. Now, do you blame high school, or do you blame modern (more visual learners) culture? It doesn't really matter. Students are different. We therefore need to take different approaches, but we have in large part to use the same faculty. We need to think creatively about how to approach this problem. It is the same pattern of thinking in terms of working with industry. A lot of faculty don't have a concept of working with industry. They are purists about academic science. We have to change the faculty and the structure in order to make these things fly.

Angelica Stacy: I didn't mean to blame the faculty. I just want the faculty to start thinking about teaching and learning. I thought I was clear that I have a constructivist philosophy, that faculty should construct their own understanding of teaching and learning. You can't do that if you already think you know everything. That is all I was saying. I want to make one point. Twenty or 30 years ago, the population of students coming to the university was the elite, upper crust. Now we should be enjoying the fact that more people are coming. I would bet that if you gave that test to the top group that is there now, compared to the 20 percent that was there 30 years ago, you would get the same results.

Timothy A. Keiderling: At our university, that is not true. Twenty or 30 years ago, the distribution was, in fact, lower than it is now.

Billy Joe Evans, University of Michigan: I would like to make a reference to Professor Eliel's comment about the master's degree. I supported him vigorously at the ACS meeting at which this was brought up. ACS will have little or no control over the Ph.D., but it can exercise a good bit of control over the master's degree. Apparently, my support did not give Professor Eliel the sense of power to really push that at the meeting. So, we failed to get it.

I want to speak to Professor Stacy's presentation. I thought it was very useful, and I learned a lot from it. My comment is this: It appears you believe that the way your students are being taught, and the way that they are learning in the paradigm you are pursuing, is superior. One way of getting around having to train graduate students as TAs is to selectively admit students from other undergraduate institutions at which this model is also promulgated.

Now, if we don't feel that this superior way of teaching and learning actually produces a better student, then this enterprise that we are in is doomed to failure, because we believe that only those students who are born with certain intellectual attributes can actually do the work that we have for them. I do not believe this is a good description of what happens to us as humans. I competed against students who would have run over me, if that were true. We can heal ourselves of physical and emotional diseases. If we are in education and do not believe that, then heaven help the parent who sends his or her child to us to have a better opportunity at living a life that is fulfilling and also contributes to the betterment of his or her fellow man.

We have been dancing around all kinds of little things, and we are smart people. We can do little things here or there, but we have not reached the core. The core of the matter is that we have made some bad hires. Universities are about teaching and learning, nothing else. You can call it research; I don't care how you define it. It is simply two things—teaching and learning. Anyone who goes into the university and is not committed to teaching and learning, learning for oneself as well as helping others to learn, does not belong there. We have a sacred commitment. We work under a charter from the nation, from the state. We get tax relief because we are given a certain job. That job is to prepare the young people of this country to be effective citizens, to give them challenging problems, to teach them how to learn to approach difficult issues. That is where research comes in. It is just another way of teaching and learning. If we are not doing that, then we should not be sitting in the chair. If we are doing that, we don't need to talk about minorities. We don't need special programs. A program that would work only for minorities probably does not work for them either.

So, it is an issue of teaching and learning. Here I am, an adult person. Am I going to run over and sacrifice a young life for this ego of mine? Faculty should be content with themselves. They should have a sense of self-worth. They don't have to publish 50 papers a year to feel as though they are worth something. This is critical. Young people are dying in our labs because of these big egos. They are little kids. They do not know. They come expecting one thing, and we give them another.

A young man who is premed asked me in class yesterday why I didn't pursue medicine. I replied that when I went to Morehouse, the most challenging and exciting discipline on campus was chemistry. So, I did chemistry. Chemistry is a good discipline for improving the lives of the people in this country. We should use it for that. Schools should be able to hire someone who says, "Look, I am good enough. I can take a little of my time, choose my problems carefully, and bring up the next generation." We have made bad hires. That is the problem. Yes, you can have academic freedom, but you don't have the freedom to ruin the young people of this country.

Dady Dadyburjor, West Virginia University: I would also like to address Professor Stacy. We are a relatively low-enrollment department of chemical engineering. So, we have the luxury of having all of our courses taught by faculty. However, that is not the point that I want to make.

Several of our faculty who are into teaching as a pedagogy have dragged me, kicking and screaming, into an advocacy of the "active learning" approach, where students sit in groups during the class time and work on assignments. The professor or the instructor walks around, sees the problems that these students are having in solving the assignment, and is able to address that directly. That is an approach that is not credited to us. Rich Felder at North Carolina State University and others have used it as well. I wonder if maybe you have thought about including the active-learning approach in any brochure or booklet that you have prepared for your TAs. I would advise it.

Angelica Stacy: Yes, and I wasn't clear about what we were trying to do. I think we want an active-learning approach in understanding the concepts. When I referred to a TA training handbook, I meant an active-learning approach to helping TAs understand about teaching and learning that adds to the handbooks that are already out there.

R. Stephen Berry, University of Chicago: I thought it would be worth pointing out that the National Research Council has just published a book that addresses these issues, which I think people in this room might find very helpful. We will try to have it available with the others that have been mentioned. This is a substantial volume called *How People Learn: Bridging Research and Practice* (www.nap.edu).[5] It came out recently. Angie knows it. I don't think there will be any problem in having copies that people can see. It also is available in full text on the Web for free.

[5]National Research Council, *How People Learn: Bridging Research and Practice* (Washington, D.C.: National Academy Press, 1999).

7

Keeping an Eye to the Future in Designing Graduate Programs

Marye Anne Fox
North Carolina State University

This chapter discusses several issues relevant to graduate education: why the federal government supports universities to conduct most basic science; how advances in information technology may affect how science is conducted; what methods ought to be considered for supporting graduate students; why research scientists incur special obligations to K-12 education as a result of the covenant with the nation; and finally, how positive the interactions between universities and private industry can be as a model for future scientific collaborations.

INTEGRATION OF BASIC RESEARCH AND GRADUATE EDUCATION

As many of you are aware, both the National Academy of Sciences and the National Research Council have an abiding interest both in the quality of scientific education and in how graduate education is successfully integrated with research in our universities. I currently serve on two committees that focus on these issues, the Committee on Science, Engineering, and Public Policy (COSEPUP) and the Committee on Undergraduate Science Education (CUSE), the second of which I chair. Both are conducting projects relevant to the issues being discussed today.

Several years ago, for example, COSEPUP produced a report that addressed the need for reshaping graduate education for scientists and engineers.[1] A concurrent report first suggested that the United States must be among the leaders, although not necessarily the undisputed world leader, in every major field of science.[2] These reports emphasized not only the importance of maintaining leadership across core disciplines, but also the need to be in a position to address interdisciplinary challenges across the entire range of human need. The latter report says that the United States must be "poised to pounce" on opportunities, whether they are initially developed here or elsewhere.

[1]National Research Council, *Reshaping the Graduate Education of Scientists and Engineers* (Washington, D.C.: National Academy Press, 1995).

[2]National Research Council, *Science, Technology, and the Federal Government: National Goals for a New Era* (Washington, D.C.: National Academy Press, 1993).

Because graduate education in the United States is intimately intertwined with basic research, national goals must include both the research objectives and the education of students within the same framework. Much of the work of CUSE seeks to leverage the quality of learning achieved by the faculty-graduate student collaboration into improved instructional opportunities at the undergraduate level. A recent publication by CUSE emphasizes the need for including the same techniques used so profitably in graduate education in the experiential learning of our best undergraduate programs.[3]

The COSEPUP report also emphasized the importance of producing well-educated Americans, people who understand the goals and achievements of science and the scientific reasoning that leads to those results. We need people who can conduct sophisticated scientific investigations, of course, but we also need those who are active in nonscientific roles in our society, especially those working in public policy, in the law, or as leaders in their communities to appreciate science. The reshaping of graduate education in science is therefore important both for the specialist and the nonspecialist, and being able to address both constituencies is an important part of academia's responsibility to our nation.

FUNDING OF BASIC RESEARCH IN UNIVERSITIES

Graduate education was not always so generously funded in the United States. Before World War II very little federal money was allocated to science in universities. It was the realization of the important contributions of science to the war effort and to the improved quality of life after the war that led to the model under which nearly all of us here today were trained. The portfolio for scientific research support has changed substantially over the last 50 years. Of course, in the early 1950s, defense was a national priority, and that emphasis continued for nearly 50 years as the Cold War was waged. But during that period, a gradual decrease in the fraction of federal support allocated to defense research was accompanied by increased payments to individuals. Beyond the support needed for national defense and mandated social payments, the discretionary portion of the federal budget is a smaller fraction of the total budget.

The same profile change over time also can be seen in the evolving budgets of federal agencies. Support for basic research from the Department of Defense has contracted while that related to human health, especially through the National Institutes of Health, has expanded. Support for fundamental science has always been a significant, but small, portion of the net federal investment.

This monetary shift is also reflected in a change in disciplinary emphasis over the same period. In 1950, engineering accounted for a large fraction of the research and development (R&D) budget, a situation that has shifted toward the life sciences over the years. The breadth of the portfolio has always reflected a cooperation between those who conduct basic research, mainly at our universities, and the federal government, while always addressing the most pressing problems faced by our society.

WHO SHOULD BE EDUCATED?

It is important to note that, despite these shifts in the nature of the work supported, the assumption that basic research would be conducted mainly at universities has been unwavering. Basic research, focusing on understanding the fundamental principles governing nature, is well aligned in the United States with higher education's principal objectives, namely, educating our students in the sciences at a

[3]National Research Council, *Science Teaching Reconsidered: A Handbook* (Washington, D.C.: National Academy Press, 1997).

level appropriate to either the scientist or the educated public. And, of course, with that opportunity comes a serious responsibility: we must accept the obligation to provide the kind of education to all interested citizens that will enable our nation (and all of her citizens) to continue to prosper. Part of the commitment made by universities in accepting support for our advanced students is that we provide the background that will prepare our graduate students to innovate in science, while contributing significantly to the education of our nonscience-major undergraduates.

In accepting these conditions as inherent in the government-university partnership, we together face certain challenges. For one, we have an obligation to provide an educated cadre who can effectively use their education to address emerging opportunities. That means neither overproducing nor underproducing highly specialized graduate-level scientists. It means not buying into a model that leads to unlimited expansion of individual disciplines, but rather encouraging teamwork in addressing socially important problems. David Goodstein, for example, reminds us that had the exponential growth in the number of physicists that took place in the 1960s continued to the present, we would now be a nation in which every citizen would hold a Ph.D. in physics. I count many physicists as among my best friends, but I must say that a nation of only physicists (or chemists, for that matter) would be a rather dull society. By including sufficient breadth within our graduate programs, we can be assured that those emerging will have the confidence to address problems not strictly confined to their own disciplines.

AMERICAN OPINION ABOUT SCIENTISTS

Surveys of American opinion indicate that scientists rank among the most highly respected professionals in the United States. But if the question is posed another way, another view emerges as well, namely, that scientists often behave as a special interest group. What are the characteristics of such a group? A special interest group resists change; it seeks additional resources as a cure for its own internal stresses; it attempts to demonstrate that members of its group deserve special treatment; it assigns its own values primacy in influencing decisions and allocations; and it claims moral superiority and privilege for its own activities. Often when we think about lobbyists, we think about big oil, big tobacco, big whatever. We should recognize that many people in this country regard big science in exactly the same category.

There is also uncertainty about whether our research, and the technology that follows from it, really does improve the quality of life. For example, the demonstrators in Seattle at the World Trade Organization meeting voiced clear concerns that technological advances are not always good. Despite our dismay that so few realize that the productivity enhancements driven by technology underlie the incredible generation of wealth over the last decade, incidents like this call into question whether we have adequately educated our nonscience majors.

Clearly, if scientists are regarded both as special interest groups and as among the most respected professionals, there is a gulf in understanding by the average American about what we do. Some, in fact, perceive us as sometimes doing pointless and trivial research. This dichotomy clearly indicates the need to communicate better with the public, especially through the representatives of the public with whom we work most closely at universities—our alumni, our students, and their parents.

SUPPORTING GRADUATE STUDENTS

What is there about the university that responds to universal human needs? What can we provide that conveys enduring value? How can we articulate a vision for an educated American citizenry? How can we ensure the integrity of our basic research effort and still provide access and opportunity to the entire range of our talented citizens? In articulating such a vision, it is useful to recall an old admonition from Hippocrates: "First, do no harm." Thus, in thinking about broad changes in the delivery of graduate education, we must be thoughtful, patient, and deliberative.

Four years ago, the National Science Board undertook a consideration of the relative merits and deficiencies of several means for providing support for graduate education.[4] We focused then on fellowships, graduate research assistantships, and traineeships. For some time graduate fellowships have been a valuable resource whereby those students judged to have the greatest potential for leadership in scientific research were identified and provided with tuition and salary support. The resulting fellowship program has produced a concentration of talented students, who in practice have chosen to pursue their degrees at a very small number of institutions. In fact, a large majority of those supported by the National Science Foundation (NSF) fellowships since the 1970s have matriculated at no more than five institutions.

One can imagine expanding the number of fellowships so that a better geographic distribution could be achieved, perhaps by allocating some fixed number to institutions that would then choose the fellowship recipients from among their own students. If we were to do so, we would, in fact, begin to contribute to building centers of excellence, geographically distributed. Such a program would also encourage domestic students to pursue advanced degrees, by providing them with secure financial support. A fellowship program would motivate students by providing personal recognition and by inspiring confidence in our most talented students in their ability to achieve within science. In addition, a program of such fellowships would relieve faculty at least partially of some responsibility for providing financial support for their students as they pursue their graduate degrees.

But with these positive features also come some concerns. In shifting to a mode in which graduate fellowships would become the primary means of supporting students, the federal contribution toward infrastructure would be seriously reduced, to the detriment of productive programs. Graduate fellowships now provide support for the student but for none of the direct costs of the research, e.g., for supplies or expendable equipment, none of the cost of state-of-the-art instrumentation, and none of the indirect costs that underwrite the university's support of the program. That could be changed, of course, but such a shift would cause major financial dislocations and would change the dynamics for strategic planning within departments and colleges.

Graduate fellowships have, in the past, produced not only a skewed geographic distribution but also a distorted demographic group. In particular, women and minorities have been underrepresented among awardees when the criteria for selection have emphasized undergraduate GPA and GRE scores. This fellowship program in many ways has faced the same challenges in achieving a diverse student population as do university undergraduate admissions procedures that emphasize high school GPA and SAT scores.

It is interesting, as an aside, to observe that there is a far better correlation between the successful completion of an undergraduate degree in science and a high school curriculum containing at least two years of a foreign language and four years of mathematics, than there is with SAT scores. Yet many universities still persist in using SAT scores as a major predictor of academic success, thereby in at least some cases disadvantaging several groups in the competition for admission and for scholarship support.

In some fields, fellowships have actually been shown to prolong the time required to complete the degree, presumably because fellowship recipients lack the financial incentive to connect early in their

[4]National Science Board, *The Federal Role in Science and Engineering Graduate and Postdoctoral Education* (Arlington, Va.: National Science Foundation, 1998).

graduate studies with a specific research group. As a result, some fellows experience weaker mentoring than would those supported on graduate research assistantships. Fellowships may also affect the quality of undergraduate instruction, by removing the best and brightest from the ranks of teaching assistants, thus depriving undergraduate students of access to the purportedly best graduate students, and denying these strong graduate students the opportunity to experience teaching at an early stage of their careers. If many, or most, domestic students were supported as fellows, a larger fraction of the teaching assistants would likely be foreign students, for whom English fluency may be a challenge. Fellowships could also cause problems in collegiality among students if they were to distort reality for some students who, like the athlete being recruited in the seventh grade, form an unrealistically positive view of their abilities.

Contrast the fellowship model with the way most graduate students are currently supported now, i.e., as graduate research assistants. Graduate research assistantships have many positive features. First, the support underlying these assistantships has been awarded to a principal investigator on the basis of a peer-reviewed proposal, meaning that the work the student will conduct is likely to be important and judged as feasible by senior scientists. The environment for conducting the work is also likely to be supportive, and the track record for the principal investigator in successfully mentoring students is likely to be strong. Because the supervisor has a vested interest in the proposed work, he or she is likely to closely monitor the progress of the student and to help him or her achieve as much as possible in as short a period as possible. Because peer-reviewed proposals are geographically distributed better than fellowships have been in the past, a focus on research assistantships would likely benefit more institutions better distributed around the country. Because mentoring by the supervising professor is aligned with the professor's self-interest, there is no need to provide additional motivation to improve the quality of graduate education.

It is true, however, that graduate research assistantships demand less creativity and intellectual independence on the part of the student. For those who are exceptionally well prepared for independent research, this can be a problem, but in my experience a large majority of first- and second-year graduate students are not yet ready for the full intellectual independence necessary to solve a significant scientific problem of their own choosing. Because the intellectual thrust of the work conducted by graduate research assistants is determined by the supervisor, there may be less personal recognition and motivation toward successful completion of the work, although the lessons learned in teamwork at an early stage of investigation may mitigate against this potential problem.

Like fellowships, graduate assistantships typically cover the costs of tuition and fees, as well as the costs of infrastructure at the sponsoring institution. Because tuition expenses can be substantial, the flexibility in a research grant can sometimes motivate faculty to take on more-qualified postdoctoral fellows rather than less-prepared graduate students, possibly to the detriment of the goals of the graduate program.

A third option for supporting graduate students is traineeships. These differ from assistantships in that the traineeship provides student support as part of a funded group project, again after peer review has established the merit of the proposed work. In a traineeship, cooperation is, by definition, strongly enhanced, and in many cases, the projects supported are highly interdisciplinary or focused on emerging new areas. Traineeships therefore can build departmental programs and camaraderie and encourage individuals involved in the collaboration to take risks in their experimental approaches more frequently than is typical with an individual research grant. Traineeships also provide a mechanism by which faculty who are between grants might access a bridge for continuity in their research programs. With competitively reviewed research traineeships, there is also likely to be a broad geographic distribution of awards.

The success of traineeships, however, depends strongly on the effectiveness of local management.

The trainee director must be rigorous in enforcing high standards for all faculty participants and must be willing to cut those parts of the collaboration that prove to be unproductive. Looser quality control in the absence of such strong management has been problematic in some programs. In addition, traineeships may provide weaker mentoring in that a supervisor may believe the student is being guided by other members of the group and not feel responsibility to closely monitor student progress.

Research in the United States is more and more frequently conducted by postdoctoral fellows, and the number and length of these fellowships are increasing. Postdoctoral fellowships have thus become an increasingly routine part of the research portfolio of universities. Because of the costs associated with tuition for graduate students, many supervisors prefer to work with postdoctoral fellows.

These fellowships unfortunately sometimes involve a prolonged period in which intellectual dependence on the research group is sustained. I loved my postdoctoral position, providing as it did the freedom to explore science unencumbered by the obligations to raise money and participate in endless committee assignments that typify academic life. But many postdoctoral fellows have rather ambiguous employment status, with some universities being conflicted about whether postdoctoral fellows are employees or students and whether they are entitled to normal employee benefits.

The conclusions reached by the National Science Board (NSB) in weighing these competing factors were that it would be dangerous to shift precipitously from the distribution of the current modes of graduate funding. Diversity in these funding mechanisms is as characteristic of the American enterprise as is the ethnic and demographic diversity that we seek.

Instead, the NSB report encouraged experimentation: traineeships for programs that encourage breadth and focus on interdisciplinary research; fellowships for some professional master's degrees; expansion of the fields for which fellowships are awarded to include nontraditional disciplines and emerging cross-disciplinary areas, perhaps supporting students who have advanced to candidacy in one discipline and wish to use their disciplinary skills to solve a scientific problem in another discipline; and fellowships that support scholarly work undertaken in industry.

ENCOURAGING UNDERREPRESENTED GROUPS

The NSB report also suggested that demographic diversity, measured by full participation of all groups, should be one of the goals of any expanded fellowship program. This is a very important recommendation. We are now at a stage in graduate education in which we are attracting far too few American citizens into graduate programs, particularly in engineering. There are many reasons for this problem, but one of the most obvious is that we fail to graduate enough Americans in the sciences from our undergraduate programs. This is especially true for women and underrepresented groups, particularly racial and ethnic minorities.

Why are minorities not graduating from undergraduate programs? Probably because they are not enrolling in undergraduate programs in sufficient numbers, or possibly because they are not prepared to succeed in these fields at the college level. There is a differential quality of teacher preparation in the precollege years across socioeconomic classes in this country. In public high schools that have fewer than 25 percent of their students taking advantage of school lunch programs—these are the most affluent of our public schools—6 percent of the teachers are uncertified in their disciplines. Science and math classes now constitute about 20 percent of courses offered in high school, and science and math represent the fields in which most of the uncertified teachers are found. Therefore, even in our most affluent schools, we can estimate roughly that 6 of 20, or roughly 30 percent, of the science and mathematics teachers are teaching outside an area for which they are prepared.

If that is not bad enough, consider those schools in which 75 percent or more of the students receive school lunches. There, 19 percent of the teachers are uncertified in the disciplines they are teaching, and if again, these teachers are assumed to be concentrated in science and math, one can estimate that 19 of 20 or 95 percent of the science and math teachers in public schools serving a poor clientele are not certified. Teacher qualification has been shown to be a major factor in the quality of learning, so one can easily imagine poor students being exposed to one, two, three, or more such teachers during the course of their high school years. Is it any surprise that students who emerge from these schools, which disproportionately include racial minorities and other underrepresented groups, are not prepared to compete on a level playing field with respect to SAT scores? If our less-affluent students are not prepared in high school, they cannot compete successfully in most undergraduate programs. And if they do not complete undergraduate programs, how can we expect them to be ready for graduate work?

RESPONSIBILITY FOR K-12 EDUCATION

Those of us who worry about graduate education must assume responsibility for improving the public schools if we are to succeed at higher levels. We cannot say this is someone else's problem. We cannot say this is the fault of the teachers, who are heroically providing their best efforts in the schools. Such an accusation would be totally inappropriate. But how many of us in the scientific community have worked with those involved in teacher training in our own universities? How many of us have encouraged our bright undergraduates who are finishing degrees in chemistry or physics or biology to even consider teaching as a career? How many of us have spoken positively to our own graduate students about considering a career in K-12 teaching?

This nation is near a crisis in elementary and secondary education. By 2010, we need two million more teachers because of anticipated retirements and the ballooning numbers of students who are already in school in the lower grades. Most of the students now enrolled in the second grade are likely to be high school seniors in 10 years, and there are 25 percent more of them now than there were 10 years ago. In my own state, North Carolina, there is a need for 80,000 more teachers over this period, and the need at the high school level is most acute in science and mathematics. The rate at which teachers are being produced by public universities in North Carolina, however, would suggest a very optimistic figure of 10,000 teachers being produced during this period.

So where are these teachers going to come from? How can we solve this problem if we do not take charge and provide positive reinforcement and incentives to consider teaching during the undergraduate years? I would encourage each of us to think about ways in which alternative certification can be accessed so that those who may have an interest in K-12 teaching have the support, moral and financial, that they need to succeed.

Let me describe one program we have initiated at North Carolina State University. It is a double magnet school to be run on our Centennial Campus (our collaborative research area) in collaboration with the local public schools. It is being constructed on our land, with students coming from every school in Wake County. The recruited students will be exposed to the best methods for teaching science and mathematics, and the admissions procedures will target talented young girls and racial minorities.

The school is also expected to draw teachers from around the state, who will learn about inquiry-based instruction and will acquire a better appreciation for the means by which students can learn more about careers in science and mathematics. Social scientists tell us that most students decide, perhaps unconsciously, during the middle school years whether they will turn toward or away from mathematics, and hence opportunities in science. A discouraging environment at that stage has been shown to turn girls away from science, and you can imagine that this discouragement would be even worse if these same students then proceed to high schools staffed by inadequately prepared teachers.

So I believe each of us, in thinking about graduate education, has a personal obligation, also to think hard about the quality of teacher training programs at our home institutions. It is, of course, quite positive to encourage our students, both graduate and undergraduate, to volunteer in K-12 classes, and I applaud those of you who today mentioned such collaborations. But our schools need more than that. It is positive to visit the local schools and to express your willingness to help. I did that several weeks ago during Teach for America week. My husband and I went to a rural school in North Carolina that had been ravaged by floods, and we taught a class about excited states—how to make a neon lamp and how it works. It was great fun and the students seemed to enjoy it.

But I have no delusions that I changed anyone's perceptions about science. A one-hour commitment is not what I am talking about. I am talking about devising means by which we can provide real assistance to these hard-working, often isolated, teachers, about helping to develop the kind of in-service training program they need. Without such intervention, our country will never develop the human resources it needs to remain among the leaders in science.

We need diversity that transcends race. We need diversity that transcends socioeconomic class. We need diversity that transcends gender. We need diversity that transcends the site at which education takes place, both urban and rural. One way we can help to attain this is to develop methods for using information technology for distant delivery of content.

TRANSFORMING EDUCATION WITH INFORMATION TECHNOLOGY

Universities are at the forefront in developing the tools that can provide long-distance mentoring as well as unprecedented access to information and knowledge. The challenge will be to develop techniques that can be used in different settings for different purposes. As technology changes, the incentives to become a scientist will be different. The usefulness of the Ph.D. experience may change; the basic research portfolio certainly will change; the sequence and incentives for responding to social needs will change; but the values of our science will persist. It is clear that universities will have to step up to the plate to fulfill the potential of the Internet.

How to address this evolving technology is a major question. Eli Noam, a professor at Columbia University, has stated, for example, that many of the current "mega-universities" are not sustainable, at least not in their current duplicative variants. Ten years from now, he suggests, a significant fraction of higher education will be offered electronically by for-profit suppliers. The significant question is whether the Internet can provide the integrated education that is characteristic of our best colleges and universities, or whether it best conveys narrowly defined training skills for specific immediate career needs.

Lest we all assume that within the next decade parents will routinely send their sons and daughters to their rooms for four years to be educated by distance learning, we should remember that Thomas Edison claimed in 1877 that the motion picture was "destined to revolutionize our educational system" and he expected that within a few years it would "supplant largely, if not entirely, the use of textbooks."

However, if we accept Internet suppliers as potential competitors for supplying at least some of the nation's needs for advanced education, we will have to state clearly what universities can uniquely provide.

UNDERGRADUATE RESEARCH

Apart from developing worthwhile programs that use the Internet, we should also think about ways that will encourage undergraduates to consider science and math careers. It is widely known that one of the most effective ways to stimulate undergraduates to continue in science and mathematics is a successful experience in undergraduate research. Why is that? It is very simple. In participating in research, the student develops a close mentoring relationship with the supervisor while simultaneously participating in inquiry-based learning. This gives him or her the power and self-confidence needed to succeed in these fields.

Let me conduct an informal poll. How many of you participated in undergraduate research? [Nearly everyone in the audience raised a hand.] Look at this. Please turn around and look at this. Undergraduate research was obviously important to this group. But how many of our institutions, as a criterion for tenure or promotion, even ask for faculty to describe their participation, or lack thereof, in undergraduate research as part of their dossiers?

If you were to suggest, at your home institution, that a worthwhile goal would be that 10 percent of your undergraduate students have a significant research experience as part of their degree work, and if you were to actively target women and underrepresented groups as part of that 10 percent, I can guarantee you that there would be a complete turnaround in the success of graduate schools in attracting American students anxious to pursue the opportunities that follow from that experience.

INDUSTRIAL INVESTMENT IN UNIVERSITIES

One way to provide that experience to an even larger fraction of our undergraduates is to collaborate with industry. The resource allocation from industry to build a cadre of active undergraduate researchers would be a win-win situation for industry and for the universities. And yet, very few of such collaborations have been formalized as undergraduate options for independent study.

The need to collaborate with industry is becoming even more compelling as federal funding outside the life sciences continues to be cut back. Public-private partnerships can address problems that are relevant and are inherently interdisciplinary. Yet relatively few such investments are being made, with typically far less than 10 percent of sponsored research on most university campuses being accomplished with industrial support.

What has caused our major industries to pull back from significant financial commitments to the universities? Clearly, the implicit assumption that such research will be funded by the federal government plays some role. In addition, some have suggested that because the results of basic research are freely published and shared broadly, rewards cannot be captured by the investor—hence, the recent evolution of more private-sector consortia to provide sponsorship and collaboration for basic research questions that underlie broad industrially relevant topics.

Although reduction to practice for key new technologies is sometimes impeded by the reduced availability of capital in tight times, basic research almost always involves the underpinning of the next generation of technologies rather than those closest to commercialization. When these challenges to developing collaborations are coupled to the rapid pace of knowledge exchange made possible through advances in information technology, industry faces a substantial challenge.

Why are these collaborations important? Unless we are able to establish workable relationships with industry, many fields, probably including chemistry, will be unable to contribute adequately to solving the social needs of Americans in the 21st century. Several things must be in place for a public-private partnership to succeed. First, you must have educated people who are driven to succeed despite obstacles and who have a firm commitment to success for themselves, their colleagues, and their communities. You must have faculty interest in and an openness to the proposed collaborations. You must have adequate facilities, which means the university must acquire land, construct buildings, and assume indebtedness for periods that exceed typical research contracts. And you must have serious partners in the venture capital community who can provide the resources to translate joint intellectual achievements into commercial successes.

It is also important that universities be flexible in handling intellectual property. Technology transfer is a contact sport, and a clear relationship must exist between the collaborative activities and those that take place on campus. These activities must be integrated smoothly into the scholarly work and reward structure of the university.

There are, of course, difficulties in doing anything new. There will be system inertia at every stage, but I am proud to say that North Carolina State University is almost unique in this country in solving each of these problems. NC State is a land-grant institution, and during its centennial year 13 years ago, it was given another land grant, a thousand acres of land in the center of Raleigh, the state capital. This land has been allocated to stimulating these partnerships while building the academic mission of the university.

There are now 54 companies that have located on the Centennial Campus. Each of these partners has pledged to conduct co-located research with one or more of our faculty, or to provide internships, co-ops, graduate fellowships, and/or research contracts. These companies range from start-ups located in our business incubator, with two or three employees, to major international corporations such as Lucent Technologies, which has chosen to move its optical networking group with 500 graduate-level employees onto our campus.

Accompanying these decisions has been an increased royalty stream that has doubled over just 2 years and an increased number of research contracts with other industrial firms anxious to tap the academic expertise of our faculty. Every day, NC State has new resources that can be reinvested in our faculty and their programs.

On the Centennial Campus, there are six research neighborhoods, each addressing a significant problem for the people of North Carolina in the areas of information technology and networking, materials science, environmental sustainability, genomic science and bioinformatics, globalization and public policy, and K-12 education. I would suggest that these six areas should be represented in each of your own stock portfolios, as they represent the future. We are proud to have established so many wide-ranging collaborations in a way that also preserves the integrity of our basic research. To the largest degree possible, we avoid conflict of interest and of commitment, but remain flexible in dealing with intellectual property. As a result, we have a unique array of public-private partnerships that both stimulate our research creativity and provide support for graduate students.

So I would invite those of you who find your way to Raleigh to make time for a visit to our Centennial Campus. It is not a research development. It is a research collaboration in which, together with private-sector partners, we have built a town—a technopolis, some say—in which the entire team focuses on the future. The basic research conducted by the university faculty is complemented by the applied research by our partners. The result is a proactive translation of technology into value for our industrial partners and a new model for top-quality collaborative research as a key option for our graduate students.

HOMEWORK

Rather than providing a summary of this talk, I'd like to suggest several homework assignments. First, please commit to explaining to your colleagues who wish to maintain support of their graduate programs how important it is to emphasize the integration of education into research and how vital it is to explain the purposes of their research to the general public. Perhaps this means giving a talk on science at your local PTA or Rotary Club. Second, please examine the research efforts of your own students and critically ask whether they are ready to work outside their specializations. Ask yourself whether you have generally resisted student initiatives to broaden their graduate experiences if it removed them from their concentration on lab research. Third, go to lunch with your colleagues who are involved in teacher education and ask them how you, as a research scientist, can help them. Fourth, take one more undergraduate student into your research group. And finally, visit one of your local industrial colleagues about the possibility of co-sponsoring a student research project.

Do remember that one individual can make a difference!

DISCUSSION

David Bergbreiter, Texas A&M University: Marye Anne, given that you have come from a state that supports research through the Advanced Technology Program and Advanced Research Program, which you are familiar with and I believe got money from, what do you think of the importance of state support of research at universities and the likelihood that other states will adopt the model that has been used in Texas?

Marye Anne Fox: The Texas initiative has two programs: an advanced research program and an advanced technology program. While I served on the Governor's Science and Technology Council, I argued for a 50 percent increase in base funding for both programs. Did you get that after I left?

David Bergbreiter, Texas A&M University: No, it is still pending.

Marye Anne Fox: Even without an increase, the program has made a huge difference in the ability of institutions, both public and private, in the state of Texas to respond to new partnership opportunities. I would say it is very important. North Carolina is far behind Texas in this particular form of investment in its universities.

Soni Oyekan, Marathon Ashland Petroleum: I am pleased to hear about a double magnet school that your university is managing. That may be one of the models for the future for helping underrepresented minorities. I have a suspicion that, for us to deal with the issues of underrepresented minorities in the chemical sciences, the universities may have to play more active roles in helping with the high school and grade school education of prospective students. I would suggest that one type of school that we would have to establish would be a boarding or preparatory school where the students are isolated for some periods from their neighborhoods so that the institutions can share intimately in the upbringing of the students. These boarding or preparatory live-in schools would provide environments for an escape from urban blight and its constant dangers for youths. They would keep the students from an abundance of poor role models in the urban streets. The boarding or preparatory schools would allow these youths to flourish in settings more conducive to their education.

Marye Anne Fox: I could not agree with you more about the importance of intervention in the middle school years to encourage interest in science and mathematics. We are delighted to have a chance to work with the public schools to do just that on our Centennial Campus, which is nonresidential. In North Carolina, a superb residential magnet school, the North Carolina School of Science and Mathematics, has been a stunning success, particularly in recruiting African-American students to science and engineering. Whether a large fraction of poor students can ever be accommodated in a residential environment is quite a challenge. Part of the problem that universities face in building a diverse student body is the court orders that impose constraints on set-aside programs. These have made it very difficult or impossible for us to have preferences or set-asides for those populations that we want to grow in order to have a broad distribution and diversity of students. In a way, I think universities were lazy in relying on set-aside programs and now, not having them available, we must be more active in promoting student interest in science and math across cultural, racial, and gender boundaries. I am pleased that North Carolina, in particular, has stepped forward to do that.

Mark Banaszak-Holl, University of Michigan: It is very exciting that you are chancellor at NC State. I was delighted to see a chemist rise to that kind of position. I am wondering if you would tell us what concrete steps you have taken in your first year to improve the reward system for faculty and graduate students at NC State to improve graduate training in the chemical sciences?

Marye Anne Fox: In the first year, we have talked about a vision at NC State. The first vision was building a diverse community that involved both demographic diversity and intellectual diversity. With respect to demographic diversity, we started a program of Chancellor's Leadership Awards, in which a quarter of a million dollars was set aside for those students who could demonstrate both financial need and a particular leadership, aimed at bringing a diverse student body to NC State University.

We have completed a capital campaign of $115 million over a target of $80 million, which is to be used for merit-based scholarships, and are seeking through new legislation about half a million dollars for support of undergraduate research. We have stimulated interdisciplinary activities by providing seed support and space, about 200,000 square feet for the six initiatives, for faculty from several colleges to work together on the Centennial Campus. We started a program that I call compact planning, which devolves decision making back to the department. In other words, what every department has been called on to do is to set goals and to accept some assessment measure that will guide the department in marching toward those goals. We have also revised the tenure and promotion procedures to enhance the importance of teaching as part of promotion. It was always true that teaching was important, but it was not the perception of the faculty that it was. I am sure that this is true at NC State, as I went through the promotion process last year with every file, and it was true at the University of Texas as well. People don't believe—this is a problem about which I would like to get some feedback—faculty don't believe that administrators are looking seriously at the quality of teaching. Because we do. We look at it very seriously.

John T. Yates, Jr., University of Pittsburgh: Your thesis extends all the way back to K-12 in analyzing the problems of graduate education in this country. I have had many connections with European graduate students and undergraduates the last few years and cannot but be impressed by the way Europeans are trained—the seriousness of purpose in high school compared to what we see in this country. Do you have any thoughts about how we might emulate the things that are happening in Europe?

Marye Anne Fox: I can say anecdotally that I have seen the same thing. A couple of years ago I lectured in East Germany before the Wall came down. The lecture, given to eighth graders, covered exactly the same material I was covering then with sophomores at the University of Texas.

There is rigor and discipline in European education that seems to be rare in U.S. high schools, and I think those are necessary qualities. But, as I noted earlier, the quality of students and how students respond are different now than in the past. Interaction with technology has changed the way our students respond, and I think that is coming to Europe as well. Globalization does not happen just economically to the textile industry or the chemical industry. It happens at every level. More and more we are going to have to develop some of the same skills that Europeans do. In particular, language fluency is something that I think will become increasingly important for our students.

Dale Poulter, University of Utah: The problem of a lack of qualified science teachers is a tremendous problem nationwide, yet there is a pool of people. This pool includes retired people and chemistry students who, late in their careers as undergraduates, decide that they might like to teach. Last week, one of my students decided not to go into teaching because of the extra two years or so it would take for teacher certification. Do you have any experience in ways to cut this knot? This is really a problem.

Marye Anne Fox: There are many people who would be interested in teaching but are unwilling to spend the additional substantial time necessary to become certified. There are several approaches we can take as a society. One is to work with the states to adopt reasonable alternate certification procedures in which the pedagogical component can be shrunk to a more reasonable level, perhaps allowing for structured supervised teaching as a substitute for education courses. It could also be accomplished by having in-class mentoring in the schools. We have a program in North Carolina called North Carolina Teach, for which state resources are made available to allow people to come from another career into the schools. It is interesting that the majority of professionals enrolling in the NC Teach Program are attorneys. Some say that they can't stand some aspects of the profession and want to get out at any cost. I wonder if that will hold up, but at least that is the original observation.

I would also draw your attention to a program that I think is a good one at the University of Texas, where it is possible to get teaching certification and a chemistry degree—and I believe this applies to physics and biology, too—within 4 years. A student must decide by the end of the sophomore year to pursue that. It would also be useful for university administrations across the country to have a fund-raising program in collaboration with the state government so that funds would be available to cover the additional expenses for staying on beyond a degree to get certification, perhaps with a payback period that could be reduced for years in service. I am working to do that in North Carolina but have not yet succeeded.

Stanley Pine, California State University, Los Angeles: Since you are connected to the National Science Board, I am asking if you can enlighten us about the rumors that there may be a new division or an expansion of higher education (I'm not sure which) in the education activities of NSF. How might that impinge on what we are doing, and how can we, as a group, help it go in a direction that we think would be good?

Marye Anne Fox: I have been off the National Science Board for 3 years, so I can't answer your question. Your information is probably more accurate than mine.

Steven Chuang, University of Akron: I wonder how we can articulate to high school students that the chemical sciences can produce more value than computer science, especially when you see that the Dow Jones dropped Chevron and Goodyear in their index?

Marye Anne Fox: How can we convince students to come into the chemical sciences when we have much stronger performance on the stock market from the dot coms and the initial public offerings and computer science than from chemistry? That is a good question. I have a lot of students in North Carolina who, after taking one course in computer science, are making a lot of money, a situation that is actually slowing down their graduation rate. An additional problem beyond what you have raised is that students are becoming less enamored of the need for credentials for them to succeed, particularly in the information technology networking arena. There will, however, always be a need for chemical scientists. In fact, chemical scientists undergird a great deal of the manufacturing strength of this country. So, while I cannot explain the stock market behavior, I think that the intellectual challenge of chemistry, its central position between biology and physics, and the fact that it will always be part of the nation's economic life have enduring value beyond the immediate creation of wealth.

Peter K. Dorhout, Colorado State University: One of the things that has come up this evening is how can we affect K-12 education. You mentioned outreach programs, many of which involve going to the schools and doing demonstrations. I think there are many other creative opportunities for having an effect in K-12 education. In the Colorado sector of the ACS, we are starting to have tutorials for K-12 teachers, not as a way of insulting their intelligence, but rather to say that if they believe they are in an area of need for polymer chemistry, or of pigmentation, or any other aspect of chemistry, we are going to have a tutorial about a particular area on a given set of days, times, and locations. These tutorials provide opportunities for teachers to ask questions and to learn about a particular area of their need. I think that is one other way we can reach out to K-12 teachers and tell them that we understand that many of them were French or English majors in college. This gives them a way to learn chemistry.

Marye Anne Fox: Peter's point is an elaboration of what I tried to mention earlier. While I think it is good for us individually and for our graduate students to have a presence in the schools, our more enduring value to the schools is as an intellectual resource. Whether that is by providing advice, seminars, or summer workshops, we have to be active and also respectful of these incredibly dedicated men and women who staff our schools. They are the backbone of our educational system and are vastly underappreciated, financially as well an intellectually.

Nicholas Snow, Seton Hall University: Throughout this meeting today, which has been outstanding, we have been discussing graduate education into the 21st century mostly at very large institutions. What do you envision as the position in the 21st century of some of the faculty at smaller research institutions and the comprehensive institutions that participate in graduate education, but perhaps not quite at the high funding and resource level of the large institutions?

Marye Anne Fox: The question has to do with the future of higher education institutions other than research-intensive universities and how they can contribute to the general development of knowledge and quality graduate education. I think the answer is likely to lie in partnerships with research-intensive universities as a means of providing equipment needed for frontier research, while simultaneously leveraging the intellectual contributions of faculty at smaller institutions. In North Carolina we have many strong collaborations with other sister institutions. And we have, as well, through some of our venture capital funds, strong connections with some of the private institutions like the ones you are mentioning. We have, for example, a start-up company in the incubator on our Centennial Campus, which is a joint partnership between NC State and Wake Forest University. It is a very effective and leveraged method by which we can both participate effectively.

Angelica Stacy, University of California, Berkeley: I have to come to the defense of my high school colleagues. Nothing is going to change until we respect them as professionals. They have a lot to offer. They have many ideas, but no time. They are in contact with students every hour of the day. There are no longer in-service days or professional days in many states. High school teachers know what they would like to do, but they have no time to do it. Yet we treat them as if we need to go and help them. In fact, I think we have much to learn from them. Until we come to the table with our high school colleagues, acknowledging them as the high-level professionals that they are, things are not going to change.

Marye Anne Fox: I hope everyone could hear Angie Stacy's statement because it is a very important one. We must be open to learning in our collaborations with teachers. Attitude is everything if we are to succeed in this important collaboration. University faculty also have an obligation to help with respect to the stream of teachers going into the schools. That is, for those prospective teachers who are still with us, we have an obligation to provide options for them so that they can be trained effectively as they go forward. We should be available to them rather than saying that we have the answers and they don't.

Angelica Stacy: And they could be available to us because they know about teaching and learning and students and diversity.

Marye Anne Fox: Absolutely.

Craig Merlic, University of California, Los Angeles: You mentioned the need for about 80,000 teachers in North Carolina. In the state of California the need is probably an order of magnitude larger. Dale Poulter had a comment on how to couple training and getting students into the pipeline for high school teaching. At UCLA we created a program in the math department that is a joint degree, a 5-year program, leading to a B.S. degree in mathematics and teaching certification. By this method we are getting people into the program early, and coupling it simplifies the process. They don't have to have a separate 2-year program. We are thinking of creating the same thing in chemistry.

Marye Anne Fox: That is a great idea. You need to work closely with schools of education and with the state certification people.

Robert L. Lichter, The Camille & Henry Dreyfus Foundation: The discussion tonight touched only peripherally on that specific issue. What that points to is the seamless nature of the entire educational process, something that most, if not all, of us understand. It is important to continue to emphasize that we are *always* engaged in education and learning, and have to engage others in them continuously.

8

The Graduate Student Perspective

Karen E.S. Phillips
Columbia University

I am a Ph.D. candidate in the middle of my fifth and final year in the chemistry department at Columbia University. This statement alone has a great deal of relevance to this workshop on graduate education and to questions that were raised during these proceedings. Columbia is a prestigious, Ivy League university that enjoys a great deal of respect from the chemical sciences community. This, of course, is one of the reasons why I decided to do my graduate studies there. Another reason is that the Ph.D. program at Columbia has a 5-year limit. Graduate students must complete their thesis requirements within 5 years, after which time they simply lose funding.

This question of whether or not there should be an enforced time limit to the chemistry Ph.D. has been raised here many times. In order for a time limit to work, a great deal of care has to be taken to ensure the student's timely progress. Columbia has devised a successful formula for this by having the students stick to a schedule of written and oral reports, poster presentations, and formal slide presentations both in and out of the area of their thesis research. In addition, all students must write and defend an original research proposal outside of their fields of specialization. There is another component to this success, however, which I think is equally important. It involves the learning and development of a certain skill on the part of the thesis advisor to guide and assess the progress of each student as an individual; to monitor the student's research more than every couple of years; and to recognize the strengths and differences of each individual. In other words, while monitoring their students' progress with the set degree requirements, advisors must also be able to broaden their criteria for evaluation of each student in order to make the most of their individual abilities. I might not be a carbon copy of all my peers, and believe that I have a unique set of skills to offer to a research program. At Columbia and in my research group, I feel that my contribution is valued.

I chose to address this issue of graduate education in the 21st century from the perspective of an organization that I co-founded a few years ago, and from the standpoint of my own specific career goals. The Columbia Chemistry Careers Committee (C^4) was born out of discussions between Spencer Dreher and myself, and began its activities in the middle of 1996. At that time, Spencer and I were both doing research with Professor Thomas Katz, and we found that our goals were similar. We both wanted to become college educators rather than university professors, and we recognized a need to educate ourselves about what would be needed to make a successful transition between graduate school and our chosen field. Although many of the types of positions that we had in mind also require a research component, the overall balance is quite different. Graduate schools, in general, provide a model for students with career paths leading toward academic environments that are similar to their own, or toward careers in industry. The needs of future small-college or community-college instructors tend to be largely ignored. It had been my experience, in fact, while visiting universities before deciding on Columbia, that teaching was seen as a four-letter word in the graduate school community. Whenever I mentioned a strong interest in teaching and asked if there were any special opportunities available to further develop my skills, I was usually told not to make this known to any professor with whom I might want to work. Somehow it was perceived that if you wanted to teach and bolster the skills necessary to become a good educator, then you could not be an effective researcher as well. This is a perception that needs to be changed.

C^4 was designed to answer some of the questions that we could not seem to ask anyone and to address more general needs as well. We recognized that we were also quite ignorant about basic job-seeking skills and the types of alternative career options that were available. We decided, therefore, that these could form a common foundation on which we could gather the support of other graduate students. Columbia University has a career services division that provides information and organizes workshops and panel discussions for prospective job seekers. After attending a few of these sessions, however, we thought that the demands of a career in chemistry were sufficiently specific to address separately.

In our department, we had the example of a number of well-respected lecturers to follow, including our thesis advisor, Professor Katz. For advice on C^4 plans and activities we sought additional help from Professors Leonard Fine and Ronald Breslow. We were lucky to garner the assistance of Joan Sberro, who was then the departmental liaison to the outside world. Joan was often the first line of contact between our department and the industrial sector or other universities. She did the groundwork for organizing all the lectures and special functions in the department. Through her we were able to get the names and contact information that we needed to start the ball rolling. With the help of a handful of other students, all with an interest in teaching, we drew up a mission statement, which we sent to the entire department. We also sent a questionnaire to the other students in the department, to find out which of the general career-based activities would interest them. The response showed that we were not alone in our ignorance.

The first activity that we organized was an interviewing workshop conducted by Sigfried Christiansen III from Smith Kline Beecham. Thirty-three students attended, a fairly large percentage of our graduate student body, which numbered about 110 at the time. In addition to learning more about the interviewing process itself and the general procedure for getting employment in industry, students were also able to ask questions about such things as the industrial working environment and what they should think about while still in graduate school if this was their chosen career path, how hiring decisions are made and what different companies might be looking for, what the outlook for a foreign graduate was like, and what seemed to be the overall state of the industrial job market.

At our résumé-writing workshop, Jim Burke from the American Chemical Society's (ACS's) Career Services Division gave a presentation on both résumé and curriculum vitae (CV) preparation to 22 graduate students and postdocs. Again, the specific needs of a chemist were addressed in great detail. This session was then followed by a series of scheduled critiques. Students were asked beforehand to bring in résumés that they had already prepared. There were time slots available for 14 individual critiques. All of these were filled, and there was a waiting list of additional students.

Through Peter Meinke, a former postdoc in Clark Still's group and the official Columbia University liaison at Merck, I was able to organize a field trip for 25 students to visit Merck's main research facility in Rahway, New Jersey. Here we were treated to a series of presentations by researchers working on a variety of projects as well as an overall tour of the facilities. Many of the students were able, for the first time, to get a true sense of what process research was all about and how this differed in the pharmaceutical industry from basic, medicinal chemistry. An agreement was made that such tours could be arranged on an annual basis or whenever there was sufficient interest on the part of our students.

As we completed these activities, we became aware of certain issues that seemed to have far-reaching significance to us as chemistry Ph.D. candidates. These issues were brought up in the context of the pharmaceutical industry. One issue is that companies generally look for employees with a breadth of experiences and knowledge. In any Ph.D. program, the tendency is to be narrowly focused on a specific research project(to become the world's expert in a certain field. This first issue really underlies the next two points that were brought to our attention as well. In industry, you need to be a team player, which is not often emphasized when you are working toward your own set of results for your own publication. The other is that obtaining a Ph.D., puts you in competition for jobs with others who have completed postdoctoral studies. Through a postdoc, one can acquire a breadth of knowledge, learn to be a team player, and develop leadership skills as well. Although we all knew people who had gained employment in the pharmaceutical industry quite easily with bachelor's degrees, it seemed that once you make up your mind to do a Ph.D., then a couple of years as a postdoc should also be factored in if you want to get a good job in industry. This may seem to be a natural and known fact at this point in my graduate career, but many students go into graduate school without this awareness.

With a broad-ranging series of seminars and requirements that are specified to be outside of your field of research, Columbia tries to do its part to add breadth to our base of knowledge. There is also a very free and open exchange of ideas and information within this department that I think is rather unusual for a university with Columbia's reputation in chemistry. Students in my department are actively recruited each year by a slew of pharmaceutical representatives. Postdocs may still be given priority, but I think that there are enough good options open to our students that others at lesser known universities might not enjoy. This is an important fact that should be kept in mind.

After these first events, I and the other students in C^4 decided that we wanted to tackle these issues from a broader perspective—not just as they relate to the pharmaceutical industry. We wanted to find out more about the job market in general, to hear more about funding issues, to hear from people who were advocates of a more interdisciplinary approach to the Ph.D. program, and to hear more about teaching. We decided to set up a forum in which these issues would be discussed. We depended on suggestions from Professors Fine and Breslow in order to identify potential speakers, but we also had a couple of people in mind whom we had been in contact with before. Starting out with a rather long list, we narrowed it down to the names with the right balance that we wanted. First of all, we were looking for people who could provide the information that we needed. We wanted also to have a somewhat controversial edge to the proceedings—we did not want everyone to agree with everyone else. Another thing that we kept in mind as we selected our speakers was finding people who would really bring attention to what we were trying to do. We wanted our forum to have some impact on the broader community of chemistry graduate schools, and we knew that one way to do this would be to have speakers whose names were well known in the community.

The title of the forum was "The Value and Future of the Chemistry Ph.D." The panelists were Janet Osteryoung, the director of the Chemistry Division of the National Science Foundation; Madeleine Jacobs, editor in chief of *Chemical & Engineering News*; Edel Wasserman, who was then a candidate for the ACS presidency; Sally Chapman, chair of the Chemistry Department at Barnard College and former chair of the ACS Committee on Professional Training; and Eduardo Macagno, dean of the

Graduate School of Arts and Sciences and associate vice president for research and graduate education at Columbia University. Ronald Breslow, then the immediate past president of the ACS, would serve as the moderator.

Once the speakers were identified and agreed to participate, then Spencer and I, with some help from other C^4 members, set out to do all the necessary groundwork for the event. We solicited funding from our department, the dean's office, the vice-provost's office, and the graduate student advisory council at Columbia University. We invited representatives from industry, and we alerted faculty and students at all the colleges and universities within our region. We arranged to have lunch with the panelists beforehand and organized a catered reception after the forum so that audience members could interact with the speakers.

One thing we quickly realized as we made all these preparations was the benefit of being a student-run organization and of having this be a student-initiated event. Not only did we get the full and immediate cooperation of our panelists and sponsors, but we also enjoyed a degree of support from our peers in the larger graduate-school community that I think would not have been there for an institutionally organized event. We had an audience of 130 to 150 people at the forum. More than half of this number were from outside Columbia University. Evidently the issues that were raised were universal enough for others to take notice and make the effort to attend.

Many of the issues that were brought up in the forum were also brought up here in this workshop on graduate education. We were interested in hearing the questions and comments of our fellow graduate students afterwards. They expressed similar concerns to the ones that we had about the current nature of the Ph.D. and how it affected the outlook for employment. They wanted to know what those in a position to change or implement policies concerning our futures had to say. They were interested in finding out about alternative career paths. They were also interested in being heard. The picture was not always as rosy as it might seem to be. Students were feeling underappreciated and overworked in some cases. There were students who just felt like cheap labor.

The forum was held on a Friday afternoon, and I remember speaking with one group of students afterwards who were from a nearby university. They thanked us for trying to raise these issues that were of such great concern to them. Then they expressed how lucky they were that their advisor had been traveling to another state for that particular weekend. Otherwise, they said, they would never have been able to come. It's easy to say that situations like this do not and should not exist, but this can also be the reality from the students' point of view. This should also be taken into account. It's easy to talk about interdisciplinary approaches to graduate education, but sometimes, unless this is institutionalized, it's not that easy for students to gain the flexibility to pursue them. It's easy for everyone to talk about the ideal Ph.D. experience, but it takes a while for these ideals to trickle down to the level where the graduate students can begin to feel the effects.

Our organization has been relatively dormant over the past two years, because we students have had too much on our hands to continue with all the work involved. We've also had some easy breaks and have been asked to co-sponsor career-based activities in the department. This was because it was recognized that students would come to support us as fellow students. C^4 started as a group of students with a strong interest in becoming good educators recognizing a need to educate themselves. I see no reason why allowances can't be made within the structure of the graduate degree for such needs to be addressed, but the student initiative should also remain as a part of these efforts. We have seen the benefits of that. Through a continued series of seminars and short workshops, we could all gain the information we need without taking too much time away from our research.

Founding and being a part of C^4 has been an invaluable experience for me. In order to bring this experience full circle, I am currently in the process of organizing another panel discussion that will focus on career options for teachers with graduate degrees. I want to have this one last activity in which the fact that I want to be a teacher will be celebrated. Again, it will take a lot of work to organize this, especially considering the fact that I am in my final year. Again, since this is not going to be provided for me, I will have to make it happen for myself. It will be worth it, though, because this would also serve as a means to pass the baton on to a younger group of students. With a little more effort, perhaps C^4 will be able to go on even after I have left. In that way it will feel as if we have made a contribution to those who will come after us. There will be some continuity, and all this effort we have made will not be wasted.

9

The Making of a Chemist: My Adventures in Graduate School

Jonathan L. Bundy
University of Maryland, College Park

As I look back on my experiences in graduate school, I think that they can truly be characterized as an adventure. My experiences have been somewhat atypical. After obtaining an undergraduate degree, I spent time in a master's program at a small liberal arts college. I was fortunate to work with an excellent mentor, who ignited my interest in mass spectrometry. Unfortunately, I had to abandon my research when my mentor's funding was cut off; my graduate program had no alternative funds available. I then had the good fortune to be able to transfer immediately to a Ph.D. program and work with one of the experts in my field. All was not settled, however, because a year later my mentor decided to move our laboratory, a daunting task in itself. On the whole, everything has worked out fortunately for me. Many graduate students, however, have bad experiences in graduate school that profoundly affect their future professional directions. If we are to better serve those who choose to pursue advanced study in chemistry, there are some issues that we must consider and some changes that must be made.

I am of the opinion that reform of the undergraduate experience is the most important thing that can be done to improve graduate education in chemistry. Obviously, a student who has an unsatisfactory experience as an undergraduate is more likely to have a bad outcome at the graduate level. There are three major areas that I think are deserving of attention at the undergraduate level—research experience, advising and mentoring, and ensuring that students have an adequate foundation in the basics.

It is fundamentally important that all undergraduates have a meaningful experience in a research setting. It is important to stress the word "meaningful" in the last statement; this should be work on a project at the bench, not just menial technical tasks or literature searching. When I was an undergraduate, experience in a research setting was a requirement for graduation. At many institutions, only the so-called "best and brightest" are given the chance to work in a research lab. If we want to increase the number of domestic students choosing to pursue graduate study, this must change so that these opportunities are open to all. Research is, after all, the cornerstone of one's career in graduate study. Those who have no experience in a research setting are more likely, in my opinion, to have a bad experience in graduate school.

Much has already been said at this workshop about the need for good advising and mentoring. I strongly agree. In the beginning, if we are to ensure an adequate supply of graduate students in the chemical sciences, we need to increase the number of students majoring in these subjects at the undergraduate level. This leads back to the need for quality instruction at the secondary school level, which often is sorely lacking. We need to make sure that those who are teaching science at this level have an adequate education in their subject area so that our youth have a positive image of science. In our colleges and universities, we need to promote the advising of undergraduates as a task to be taken seriously by faculty. Advisors need to take an active role so that their advisees are adequately informed about graduate study.

We must also be sure that those we send to graduate school are adequately prepared for the rigors of graduate study. This includes education in the basics of work in the lab. All too often I have seen new graduate students who cannot properly prepare a solution, or worse yet, do a simple mole conversion! Undergraduate laboratory instruction should stress these basic skills at all levels. The use of "open-ended" laboratory experiences, which have recently come into vogue, should also improve the preparation of students for graduate work. After all, in a research laboratory, there is no lab manual that will magically produce results if diligently followed.

Once students get to graduate school, one of the initial experiences that they usually have is being put in charge of several undergraduate laboratory sections as a teaching assistant (TA), often with minimal training. Unlike the humanities and social sciences, graduate students usually are not given the opportunity to lead a lecture class. TA experience in laboratories often does not provide adequate preparation of those who ultimately may want to pursue a career as a faculty member at an undergraduate institution. For those who want these experiences, we should provide avenues to obtain them, such as "teaching fellow" or "faculty apprentice" programs.

Another issue is the situation at larger institutions where a large number of TAs are required to support the large enrollment in undergraduate chemistry courses. There is an impression in the chemical community that the pressure to staff lab sections may lower the standards for admission to a graduate program. This obviously is a disservice to both the undergraduates receiving instruction from a less-than-adequately prepared TA and to the graduate students who may not adequately be prepared for graduate school. For those institutions requiring a large amount of TA labor, alternative sources of teaching assistants should be pursued, such as retired chemists (or science teachers) and recent college graduates desiring a break before going to graduate school.

As I have mentioned before, the cornerstone of one's experience in graduate school is research. There has been some discussion as to sources of funding for students actively pursuing research. Traditionally, this has been through research assistantship funds directly awarded to faculty members through grants. A younger, less-established faculty member often does not have sufficient funding to directly support students, forcing them to work as a TA beyond the required period. An alternative idea is to give fellowship money directly to students, as is often done in the biomedical sciences. I think this is an excellent idea. It might enable younger, less-established faculty to be able to attract better students instead of choosing from the "bottom of the barrel." In addition, students would be able to make the critical choice of advisor without having to place funding as high on the list, as many students often do. Industrial partnerships also are an alternative source for funding graduate students, and I think an excellent one for those who are definitely set on a specific industrial career. However, they seem to be a "flavor of the month" as of late. It is important that these programs do not sacrifice the integrity of graduate training for industrial interests. With the diverse job prospects for those trained in the chemical sciences today, students who are planning on a nonacademic career can still receive adequate training in a nonindustrial graduate program.

The relationship between a student and advisor is also important. Often students feel that they are just a cheap source of labor for their advisor to exploit as needed. It is critical for a student's professional development to be involved in decision making as to the direction of his or her research, instead of being treated as a glorified technician. This brings up the issue of how long is long enough for the completion of a Ph.D. I think it should be of a sufficient period to allow a student to provide a significant contribution to their particular field of the chemical sciences. Although it may be tempting to "squeeze" one more compound or paper out of a student, advisors must be careful not to exploit students by extending their program beyond a reasonable time.

To conclude, I feel that our system of making chemists is an excellent one overall. After all, our graduate programs are the envy of the world. Several individuals have mentioned at this workshop that we should not fix what is not broken. I think this is an excellent idea. We do however, need to ensure that we have a supply of "raw material" to continue our excellence. The suggestions that I have mentioned concerning education at the undergraduate level need to be heeded if this supply is to be sustained. We also need to ensure that the concerns raised by graduate students are treated seriously and thoughtfully, so that the majority of those who pursue graduate study in the chemical sciences graduate from their programs with the impression that getting an advanced degree was worth it.

10

A Perspective from a Former Graduate Student

Judson L. Haynes III
Procter & Gamble

In my graduate education experience, I have come to the conclusion that graduate school is the modern-day version of the mythical labyrinth of Crete, especially for underrepresented minorities. Often, minority students find that there are many unwritten expectations, a lack of a support network, and minotaurs to encounter along their journey. Some make it out. Others have bad experiences, get consumed with the maze, and never make it out.

As a recent graduate of Louisiana State University (LSU), I would like to give my perspective on graduate school, the chemical industry, and chemical education. My career as a chemist began at an early age in my grandmother's kitchen making soap, probably my first organic experiment. This really sparked my fundamental interest in chemistry. I was fortunate, because my passion for chemistry bloomed while I was in high school. I had a gifted high school chemistry teacher, who also served as the physics teacher. He relayed his passion for chemistry and chemical engineering to me, thereby nurturing my growing interest in these areas.

One thing that really impressed me about him was that one of his former students had gone to the state university and graduated first in his class. More important, this student was graduating with a degree in chemistry and was going to medical school. So, it excited me to know that I had the same teacher. I was sitting in the same class and getting the same notes that this individual had received in his education, and he went on to do bigger and better things.

The real turning point in my education began when I entered my undergraduate university. Hampton University is a small, historically black college, located in Hampton, Virginia. At the time, I didn't know that Hampton had a very strong (American Chemical Society [ACS] certified) program in chemistry. Chemistry was one of the most demanding majors on campus. When I arrived at Hampton and got settled, I started getting into classwork and had the opportunity to participate in the Minority Access to Research Careers (MARC) program.

The MARC program is a federally funded program that provides scholarship money, tuition, and summer opportunities to minority students in science at historically black colleges and universities (HBCUs). The most beneficial part of the MARC program (a taxpayer-funded program) was that every summer MARC students would go to different universities with different research programs and conduct undergraduate research. This was great for me because every summer I had something to do. I did not have to go out and try to find a job. I was actively participating in my career as a chemist from day one. These experiences were invaluable because they let me learn what chemistry was about. I worked for people who were easy to get along with and others who weren't so easy. However, I realized that the bad experiences were isolated. Ultimately, the MARC program taught me how to conduct research.

I was able to formulate my own theory about graduate school from my undergraduate experiences. For instance, I had one experience during a workshop at Purdue University in the summer of 1992 that focused on careers in chemistry. This workshop gave freshmen and sophomore chemistry majors from various universities across the country the opportunity to go to Purdue University for a weekend to talk with professors and meet with graduate students to find out what chemistry was about. I was able to meet a lot of people and got some in-depth knowledge about graduate school while I was still an undergraduate.

I would like to list the summer programs that I attended to show the experiences I was able to benefit from because of the MARC program. I spent one summer in Hampton and a summer at Virginia Tech in the National Science Foundation Summer Research Program. That was a very valuable experience. Following graduation, I spent the summer in the Washington, D.C./Maryland area at the National Institutes of Health, again funded through the MARC program.

I would like to pass on some advice to undergraduate students: summer programs are very valuable. Summer programs give you the opportunity to see what a profession is like. If you really want to go to graduate school, it gives you the opportunity to see what it is like, rather than waiting until you have enrolled in a graduate school and then finding out you have been matched to the wrong university.

The application process from undergraduate to graduate school was like a maze. Nobody I knew had a clear and direct route on how to enroll in or select a graduate school. So, I chose the same strategy I did as an undergraduate. I applied to a variety of schools, but I made my selection based on what field I was interested in and who the leaders I was interested in working for were within that field.

After deciding where to apply, I immediately contacted the various universities and professors at the same time I applied. So, by the time my application arrived at the university, I knew the people there and I knew the people I would like to work with in the department.

One thing that I tell students is that it is important to select your university based on your needs. For example, a student might decide to try to go to Harvard, which is a very prestigious school. However, if students want to pursue analytical chemistry, it won't make any sense to apply there because Harvard doesn't have an analytical program. If they did choose to go there with the intent of becoming an analytical chemist, it would be a mismatch and they would likely have a bad experience.

I have run into a lot of graduate students who have had that experience. So, as I say to students, try to find a university that is going to be a good match for you. Try to identify leaders in your field or your career interest that will help you get to where you need to be or want to be. For those who are undecided about where they want to go, it is important for them to find an environment or a university that will provide support and education according to their needs.

Prior to my graduate career, I used my strategy and selected Isiah Warner as the person I wanted as my graduate adviser. I had tracked Dr. Warner's move from Emory University to Louisiana State University during my junior year in college. I knew about his research and was highly interested in chromatography and spectroscopy, which, as a matter of fact, was the basis of a lot of my foundation coming out of Hampton. Most people probably choose their graduate schools based on location of the school. My outlook was to go where he was, because I knew that I needed to go to that school and get the skills that I needed to be where I wanted. If Dr. Warner had gone to the University of Alaska, I would have gone to the University of Alaska. This is a very important thing for students to realize: if they want to pursue their dreams, they might have to go and chase them. They can't just stay in their little box or their little corner of the world and expect their dreams to come true.

Graduate school was a new environment, but I was ready for it because of my previous experiences. I knew a lot of graduate students when I was an undergraduate, so I was pretty well prepared for it. When I came into graduate school, I was a typical graduate student in some ways, scared and worried that I didn't know anything. All I knew is that we had to pass these cumulative exams. This was a major issue. I went in with the expectation that I just wanted to pass one cumulative exam in my first year. I would have called that a victory.

I learned immediately, once I got to LSU, that I had to study very hard. I studied with some of my friends, including one of my colleagues, Victor Vandell. Victor and I had a competition to finish the cumulative exams. Victor is an organic chemist; I am an analytical chemist. We both took cumulative exams at the same time and finished them within the first year. That was a major victory for me, especially since my expectation coming in the door was to pass just one within the first year.

After my first year, the mythical minotaur appeared. The students at the university believed that minority students were bringing the standards down. Need I remind you that the cumulative exams were blind. You don't write your name on them. You write your social security number, thus eliminating bias by the grader. In most cases, the cumulative exam topics were as broad as analytical chemistry or organic chemistry, or more focused topics such as mass spectrometry or know all the A-page articles in analytical chemistry for the last 5 years. I don't know what gave the students the impression that the minority students had some distinct advantage, that we were somehow getting over, and we were lowering the standards. No, the minority students were not lowering the standards. If anything, we were pushing the standards higher, because we came in the door and we studied and worked diligently. And it paid off.

A lot of reeducation had to go on in the department as was seen from this experience. Many students had never heard of historically black colleges and universities. Thus, they assumed that the HBCU was an environment where people party for four years and then ended up in the same graduate school. That was not fair. But a lot of them didn't know that the HBCUs had ACS-certified programs. That means we took the same courses as someone who goes to any certified school. Our degrees were ACS certified, just like our peers. We were prepared just as they were. This was an important reeducation for everyone.

I would now like to pinpoint some of the things that made my career in graduate school instrumental to going into industry. The first thing was the development of oral communication skills. We did a lot of presentations. Dr. Warner was very demanding of us. Our group meetings were on Saturday mornings at 8:00 a.m. After three or four years of this, I enjoyed getting up and working on these issues and details early in the morning. I found out that you could do a lot in the morning, especially in a research laboratory that is usually overcrowded.

Another thing that was instrumental was presentation skills developed at national conferences. My first major presentation was at Pittcon in 1995. I spoke in an auditorium that seated 500 individuals. I spoke prior to one of Richard Zare's students, so everybody was in the auditorium. This was my first talk, and I was saying to myself: "There are 500 scientists in here and I am going to address them." Well, a frightening experience turned out to be a valuable experience. It allowed me to communicate my research ideas and my research work to a larger audience. I got a lot of feedback from that, and I was grateful to have that experience.

Also instrumental and very important is the communication skill of writing. We were urged to continuously write up our work and submit it for peer-reviewed publication. We got bashed often, but we learned a lot about writing papers and doing research. This is instrumental, because in academia publish or perish seems to be the theme. If you don't prepare to write in graduate school, you are going to have a difficult time when you get out because your expectations are different from that of the discipline of chemistry.

Another important asset in graduate school, similar to undergraduate school, was the availability of instrumentation, which at LSU was incredible. Even my undergraduate university, which was funded mostly by national funding councils, had ample instrumentation. For instance, at Hampton we had spectroscopy instrumentation. At LSU, we had a better grade of spectroscopy instrumentation.

It was important throughout graduate school to learn how to use and repair instruments that are state of the art. This allowed us to go to conferences and talk with vendors who were selling state-of-the-art instruments. This was especially helpful because my professor liked to buy instruments without a service contract. Imagine toiling away in the lab for years and the instrument breaks down when you are getting critical data. We had to learn how to fix all of our instruments. We often complained a lot about this, but it was an important skill. It improved our problem-solving skills, in addition to applying problem-solving skills to our research. It was important to know how the instrument worked as well as how to repair it.

I would like to give a few suggestions from my experience and from talking with some of my colleagues about faculty and funding agencies. Professors are busy people. We all know this. We forget, however, that students imitate the behavior of their professors. If the professor doesn't come to seminar, students think they do not have to come to seminar. If the professors don't show up, students think they do not have to show up. They believe that the most important requirement is to get the professors' research done by any means necessary.

Graduate school education is more than just research. Part of graduate school is interacting with people from various countries and different nationalities. It's also learning from different people with different aspects. There are clearly a lot of things that go on in graduate school, but I would like to ask the faculty to keep in mind that the attitudes you project are important. Often, the students try to project those same attitudes without realizing they might not have the stature of the faculty.

I want to tell the funding agencies how great it is that funding is available. I am a product of taxpayer money. Taxpayers paid for my undergraduate education. They paid for my graduate education. They paid for my summer education. This is very valuable. I learned early that if I dug in and really got involved, I could spend time advancing my career and learning at the same time. I never had to worry about money issues. There was constant funding. I knew that as long as I performed, the funding would be available; I would rather perform in a chemistry laboratory than work at McDonald's flipping hamburgers for the summer. I would encourage the funding agencies to continue funding graduate students and to continue funding outlets so that graduate students can grow.

I would like to introduce a new idea to funding agencies—diversity. I guarantee that if funding agencies explicitly say that diversity should be a part of research programs that you would see universities and departments currently resistant to this change begin to make changes.

I would like to briefly talk about my transition into industry. For me, the transition felt strange because I had never really had a job. All of my experience was in research. Here I am, 27 years old, a full-grown adult, accepting a position at Procter & Gamble. This was my first job. I have been in school all my life. My transition from graduate school to Procter & Gamble was very smooth. At Procter & Gamble, I will have responsibilities for managing people, working in teams, and communicating. These were three skills that were constantly reinforced throughout graduate school, which emphasized learning how to communicate with other people, communicating your results and your ideas, both verbally and in writing, working in teams, and also just learning how to get along with everyone.

The minotaurs will always be there. The problems and issues that conquer students will always be there, and the maze will always be there. Students don't necessarily want a map of how to get through the maze. The most important thing is to get students involved early and often and keep them motivated.

11

New Students, New Faculty, and New Opportunities: Preparing Future Faculty

Richard A. Weibl
Association of American Colleges and Universities

I would like to start with a short survey. All of you can do this. Those of you who work in industry may have to make some translations. When I ask about work in colleges or universities, think about your company. When I talk about faculty roles, think about the kinds of activities that you do in your own setting. This quiz invites you to recall your graduate school years. The questions are based on the actual work of faculty members at most of the colleges and universities in the United States. The items should be familiar to many of you.

Reflecting on your graduate programs, did you have a course, seminar, workshop, or any formal preparation for the following:

- **The Teaching Roles of Faculty**

  —Learning about theories or research literature regarding how students learn?
  —Setting learning goals and assessing student performance?
  —Understanding the meaning of and differences between a "liberal," "general," or "professional" education?
  —Conceiving, designing, implementing, and evaluating a course?
  —Linking one course to another, a sequence, or a program?
  —Integrating research activities into classroom activities?
  —Designing courses for nonmajors?

- **The Research Roles of Faculty**

  —The history and meaning of "academic freedom"?
  —Professional ethics, responsibility, and conduct?
    - Understanding intellectual property rights?
    - Developing a line of research?

- Involving students in research activities?
- Integrating research activities into classroom activities?
- Writing proposals for funding?

- **The Service Roles of Faculty**

—Review, promotion, and tenure processes?
—Roles and responsibilities for institutional governance?
—Effective mentoring of undergraduate and graduate students and junior faculty?
—Reviewing academic programs for integrity, cohesion, and significance?
—Serving on committees, with faculty peers from other fields, in service to institutional goals?

In reviewing this list, some might have recalled conversations among fellow graduate students, maybe even with a faculty adviser. In some cases you may recall advice that too many graduate students today can repeat verbatim, "Don't waste your time with those things." I suspect many of you had no formal preparation for most of the tasks you would be asked to do in your first year as a faculty member.

Why should doctoral students, the future faculty, learn these activities? Should graduate programs be expected to teach them? We think so. That is the position of the Preparing Future Faculty (PFF) program, sponsored by the Council of Graduate Schools and the Association of American Colleges and Universities with financial support from the Pew Charitable Trusts and the National Science Foundation.

We clearly are saying, "Yes, it should be a part of the way future faculty are prepared for the full range of faculty roles and responsibilities." This assertion is based on several assumptions:

- The Ph.D. is, and should remain, a research degree.
- Not all doctoral students aspire to faculty careers.
- Not all doctoral programs aspire to prepare students for faculty careers.
- Too many doctoral programs are failing to provide students educational experiences that make them "job ready," whether for industrial or academic work, upon graduation.

PFF programs are not to be confused with teaching assistant (TA) development programs. TA development programs have made great improvements in the quality of basic preparation available to graduate students with teaching assistantships. PFF programs often build upon this foundation, giving graduate students a more complete awareness of the work faculty members do in each of their roles—teaching, research, and service.

A recent issue of *Chemical & Engineering News* (November 15, 1999) features a special report on the employment outlook for 2000. The articles make it clear that graduates will find employment, but that skills and interests beyond the research paradigm are being sought by a variety of employers. When department chairs at major research universities assert[1] that candidates "should be both qualified and interested in becoming educators," have "a creative interest in undergraduate education," and "definitely interested in interdisciplinary areas," we must acknowledge that something beyond the traditional research experience is desired in a "job ready" candidate.

[1]"Demand 2000," *Chemical & Engineering News*, November 15, 1999, pp. 38-46.

KEY CONCEPTS OF PFF PROGRAMS

The key concepts that guide our program for graduate students, the future faculty, are these:

- **Graduate students should learn about the academic profession and have experience with the kinds of institutions that may become their professional homes.** The majority of the doctoral students who seek faculty positions will not find employment in research universities. Doctoral programs should provide students the opportunity to make informed choices about the variety of academic career options and work settings in higher education. PFF programs create institutional and departmental partnerships, forming a collaborative network of faculty and academic administrators committed to the future professoriate.

- **The graduate experience should equip future faculty for the changes taking place in teaching and learning so they are adequately prepared for the classrooms of tomorrow.** Yesterday, Professor Stacy introduced us to a constructivist point of view, something relatively new in thinking about teaching and learning. These ideas about effective teaching apply in both the academic and industrial work settings. PFF contends that doctoral programs should provide students the opportunity to learn about and acquire the best skills to teach and communicate with others.

- **Graduate programs should include formal systems for mentoring in all aspects of professional development. Let's not make the assumption that faculty know how to mentor.** Most were never trained to be a mentor, and those who are most successful accomplish a great deal without formal training in this area. We all know many of us could have used some help in learning how to be a better mentor. Syracuse University has a formal program that most of their faculty have now completed in order to be "certified" as a faculty mentor on its campus and to be rewarded for successful mentoring in the promotion and tenure process.

- **Graduate programs should include opportunities for students to develop professional expertise in teaching, research, and service, as well as opportunities to learn how to balance and integrate these responsibilities.** Each of us struggles to find a balance between the many demands competing for our most valuable resource—our time and attention. Graduate students will tell you that they, too, have this struggle. Yet when programs create artificial boundaries or expectations (for example, working 60 hours per week in the lab), students are not given opportunities to learn about other roles or how to balance the roles.

- **Apprentice teaching, research, and service experiences should be planned developmentally so that they are appropriate to the student's stage of development and progress to degree.** Urban legends tell of doctoral students being assigned to teach classes, lead labs, or serve committees without the barest of preparation. And while many might believe this "swim or sink" approach is effective, I suspect many of us who have had that experience are convinced that there must be a better way. How could you do it differently?

- **PFF experiences should be thoughtfully integrated into the academic program and sequence of degree requirements.** Program review and design are a very important collective responsibility for university faculty. We all share a concern about time to degree and recognize that each new idea cannot simply be added to the existing program requirements.

CAMPUS-BASED PFF ACTIVITIES

In 1993 the Association of American Colleges and Universities (AAC&U) and the Council of Graduate Schools (CGS) invited 17 graduate schools to design and develop model PFF programs based on the previously mentioned key concepts. Since 1993 more than 200 colleges and universities have been involving hundreds of faculty and thousands of graduate students in activities including course work, weekly professional seminar series, campus visits, participation in teaching, research, and service activities with mentors, and more.

For example, university-wide PFF programs are

- Offering a seminar that addresses general issues in college teaching;
- Organizing a seminar that addresses professional and career issues (e.g., professional ethics, academic freedom, intellectual property, and promotion and tenure processes) and is led by faculty and academic administrators from diverse institutions;
- Discussing teaching in a multicultural setting and teaching about diversity;
- Offering certificate programs in PFF and noting accomplishments on student transcripts;
- Explaining academic governance systems and inviting graduate students to attend faculty meetings or committee meetings;
- Helping students develop portfolios documenting expertise in teaching, research, and service; and
- Training faculty to mentor.

Departments participating in PFF are

- Inviting doctoral alumni to discuss how their careers do or do not connect with what they did in their graduate programs;
- Encouraging the development of professional portfolios;
- Offering courses on teaching in their respective disciplines; and
- Creating forums to discuss faculty histories, career paths, and alternative professional lifestyles.

College and university partners are

- Discussing their distinctive academic missions and different academic cultures;
- Assigning PFF students a faculty mentor;
- Inviting PFF students to join them in supervised teaching of course units or entire courses and providing feedback;
- Inviting PFF students to attend faculty, committee, and departmental meetings and to later discuss their interpretations; and
- Arranging opportunities for PFF students to discuss the graduate school experience with undergraduates.

Participants in the national program receive a dossier letter that says they have participated in our national program and that urges hiring institutions and departments to pay special attention to the candidate because he or she worked above and beyond to be prepared for a faculty role.

PFF IN THE CHEMICAL SCIENCES

The national PFF program, directed by AAC&U and CGS, has several funded initiatives to develop and institutionalize model programs. The National Science Foundation is supporting working with five disciplinary societies (American Association of Physics Teachers, American Chemical Society, Special Interest Group on Computer Science Education of the Association for Computing Machinery, and both the American Mathematical Society and the Mathematical Association of America). Each society has conducted a national competition and selected four departments to develop and implement PFF programs in their departments.

The Education Division at the American Chemical Society did an excellent job in recruiting innovative proposals, so much so that we in the national PFF office shifted funding so that one additional chemistry department might join the program. The chemical science participants include the following:

- **Duquesne University,** Department of Chemistry and Biochemistry, working in partnership with Chatham College, Community College of Allegheny County, La Roche College, Seton Hill College, St. Vincent's College, and Thiel College. Contact: David Seybert (seybert@duq.edu);

- **City University of New York-Queens College,** Department of Chemistry and Biochemistry, working in partnership with Queensborough Community College, Baruch College, and Manhattan College. Contact: Thomas Strekas (thomas_strekas@qc.edu);

- **University of California, Los Angeles,** Department of Chemistry and Biochemistry, working in partnership with California State University, Fullerton, Mount St. Mary's College, and Mount San Antonio College. Contact: Arlene Russell (russell@chem.ucla.edu);

- **University of Massachusetts-Amherst,** Department of Chemistry, working in partnership with Amherst College, Hampshire College, Greenfield Community College, Holyoke Community College, and Smith College. Contact: Julian Tyson (tyson@chem.umass.edu); and

- **University of Michigan,** Department of Chemistry, working in partnership with Baldwin Wallace University, Calvin College, Eastern Michigan University, Grand Valley State University, Hillsdale College, Hope College, Kalamazoo College, Oakland University, and Oberlin College. Contact: Brian Coppola (bcoppola@umich.edu).

Present with us today is Professor Chris Bauer from the University of New Hampshire, whose institution participates in the PFF program funded by the Pew Charitable Trusts. They are doing marvelous things at New Hampshire, including offering the field-based course "Teaching and Learning in Chemistry," using PFF activities to recruit doctoral students to their chemistry programs, and offering special fellowships supported by the Camille and Henry Dreyfus Foundation to PFF students.

You can learn more about all the PFF programs by visiting our Web site at <www.preparing-faculty.org> or by sending e-mail to pff@aacu.nw.dc.us.

My closing thought was suggested by a faculty partner in the University of Minnesota cluster: If we consider that PFF is a radical experiment in dispersing the ownership of the preparation of future faculty across a wider collaborative network, we should stress that PFF also represents a new way of thinking about the professoriate as a community of scholar/teachers embracing the full range of academic cultures.

Panel Discussion

Stanley Pine, California State University, Los Angeles: I really appreciate the organizers getting these students here. It has been a very important part of this conference. One of the themes that came through, particularly from the students who spoke this morning, was a lore of graduate school that an advisor/boss will not let the students do anything outside of the lab. I think part of it is just something students always say, but there is some truth to it, and I hope one of the things that we can work on in our thinking about the graduate education program is how to get beyond that. Clearly, as we have talked about in this session, there is more to do than research in the lab. The fact is that there are some students who are a little afraid that if they spend any time outside of the lab broadening themselves they are going to be in trouble. Do the students have any comments on this?

Karen E.S. Phillips, Columbia University: First of all, I don't want to give the wrong impression. I was able to do all of these things with C^4 thanks to the cooperation of my advisor. He was able to look the other way and have different expectations of me during that time, and I think that he appreciates me for the strengths I have and the things I have to offer that might be unique. It is generally a problem, however. We would have had far greater participation from other students if this were not the case. And, as I said, students from other schools were coming to us and telling us that it is so hard for them to get away to do anything. So it is a reality, even though I know that I have been given a lot of flexibility.

Laren Tolbert, Georgia Institute of Technology: I have faculty members who literally tell their students not to go to a seminar because they don't want them out of the lab, even for something that is obviously part of their education. Karen demonstrated that graduate students have a lot more power than they give themselves credit for making things happen, and this is important. But I also have a concern. I want to know how you deal with it and what will happen when you leave. Who takes over and how do you make sure this continues?

Karen E.S. Phillips: You see, this is another issue. Since we put on the forum, I, for example, had to concentrate on my original research proposal. Everyone in C^4 got really caught up and busy with their degree requirements. So we have not been as active since then, simply because we haven't had enough time. I am trying to do things to revitalize the organization before I leave in the hope that it will be passed on to future graduate students and students that have come in since we have had these activities. That is why I am really, really anxious to put on this one last event and, again, make it an event of particular relevance to me. But the fact is that it is hard to keep the ball rolling simply because the students don't have much time.

Again, what I tried to hint at before is that if the institution or organizations could, for example, provide us with the funding to do these things, that would be one less thing that we would have to concentrate on, one less thing that we would have to spend time doing, if funding were already in place. But I still think that it is important for the activities to be initiated by students in order to get cooperation from everyone involved.

Soni Oyekan, Marathon Ashland Petroleum: I want to applaud all the speakers this morning—Karen Phillips, Judson Haynes, Jonathan Bundy, and Richard Weibl. They touched on many aspects that also lead to some of the things that we have been hearing throughout this workshop. Karen and Judson emphasized the diversity of thoughts, actions, and programs, and the impact on their lives. Jonathan spoke of the need to have an adequate supply of students coming through the pipeline. He stopped at the undergraduate stage, but as Chancellor Fox said last night, we must go back all the way to the elementary schools to make sure that we have students to fill up those slots that we want in the future.

In addition, while doing that, we cannot forget the underrepresented segments of our society. I must emphasize that again. I think it is important for us to look at countries that have failed to do this. For example, in my native country, Nigeria, the funds were not spent on education, and what you have today is a lot of panicking, a lot of unemployment, and weapons in the wrong hands. As a result there is no comfort or safety for anybody, including the wealthy. In this society, we can see what we gain from people when we give them the opportunities and ensure that most of our population segments are fully employed. We have seen that in the discussions in this meeting. I would like to remind this great body that this nation leads in terms of education. We need to put in place a national plan, so that 30 or 40 years from now we don't fill up our prisons and cause difficulties for our society. It is more than just getting some superlative people to go into teaching. It is making sure that our population is fully employed.

Lynmarie Posey, Michigan State University: We have heard a lot of discussion over the past day and a half about trying to broaden the background of graduate students, while maintaining some control over the length of the Ph.D. program. I would like to get the graduate students' perspective on the best way to achieve this goal. What do you feel is the best way to broaden your background? Is it through course work, for example, by increasing the number of courses required or instituting rules that require students to take courses in areas beyond their area of specialization? Is it through the research experience? Or do you think that broadening students' backgrounds is best achieved by informal experiences, such as forums to explore career choices?

Judson L. Haynes III, Procter & Gamble: I think one good way to broaden the graduate student perspective is by having affiliations with or working within the professional organizations, such as the American Chemical Society (ACS). It is incumbent upon the students to start early, get involved, and get to know people in the ACS and network. If students don't, and if they just spend 5 years in the lab, it's possible that by the time they start interviewing they can be socially maladjusted. They have had little to no personal interaction with people outside of their research, which is different from industry, where chemistry is often a teamwork effort. So, if you have toiled in the lab for 5 years and you come out with no interpersonal skills, you might be able to get the job done, but nobody is going to like you. It is important for students to start getting involved early with committees and with their local sector of the ACS. ACS is everywhere. It is something students have to do.

Karen E.S. Phillips: Yes, but I would add that, as you mentioned, having forums where we can get, as Richard mentioned, the alumni to come in and talk to us about their careers is really important. It doesn't necessarily have to be a full course. It can simply be anything that puts us in touch with the outside world and addresses how graduate school helped people to make the transition.

Lynmarie Posey: I think the faculty response is to add courses, which then potentially increases the amount of time that students spend in graduate school. That is also a major concern for junior faculty, because they would like to get students doing things in the lab as soon as possible.

Richard A. Weibl, Association of American Colleges and Universities: I would like to add that if it is important to happen, it must happen as a part of the degree program. That's the bottom line. If you think it needs to happen, make it happen intentionally within your program structure. If you create a program structure that forces students to be in the lab 80 to 90 hours a week—I recall the article in *Chemical & Engineering News* about graduate student stress leading to suicide—that is not going to give a student a broad experience. And if you are creating an educational context—and I assume it is an educational context and not a cheap labor one—then you have to look at the program and decide where to make room and sanction student participation in these kinds of activities. Also, we must determine how we will dedicate resources so that the students can do these things. If you don't do that, then it is extracurricular activity, and junior faculty, as well as senior faculty, will not value it. That is the only way it is going to happen.

Jonathan L. Bundy, University of Maryland at College Park: I think the most important thing, as he said, is actually getting out of the laboratory, going to the national meetings, and interacting with your fellow graduate students both inside and outside of your laboratory. We are lucky to have a lot of postdocs from diverse backgrounds to help give us a perspective on many different things. It is important for graduate students to get out to national meetings, see what the other half is doing, and broaden their horizons from the black box of their research.

Karen E.S. Phillips: Yes, but how many people are actually encouraged to do so?

James D. Martin, North Carolina State University: I would like to make a comment with respect to today's discussion of preparing students for the professoriate. We need to give some consideration to this in the context of comments made by Steve Berry yesterday on the institutionalization of interdisciplinarity. It is extremely important to give exposure and encourage people to gain skills in some of these areas, but institutionalizing requirements for professoriate training can run a bit amok.

At North Carolina State we have a professoriate program in which I participated with one of my graduate students. I would argue that the institutional aspects of the program were moderate to okay. However, the real value that came out of the program was the time that I spent with the student back in my office discussing the realities of the professoriate and educational philosophy, as well as the mentored experience in the classroom. It was the personal interaction, the mentoring, as opposed to the formal instruction that had the greatest impact.

So, when I looked at Richard Weibl's "did we have formal training in graduate school" list, my response is "absolutely not" to formal instruction, but I can say "yes" to experience or training in almost every one of those areas through interaction with my mentor and other chemistry faculty. We need to encourage ourselves as mentors to be available to our students and encourage our students to take action, more than creating a new set of requirements. Yes, we are busy, but I keep my office door open. If you come to my office and ask me if I am busy, the answer will be "yes." But I tell my students to grab me by the neck if necessary and sit me down and talk to me.

If we want to improve the preparation of students, I believe we must focus on our availability and interaction with them, not institutionalization of a new program. We don't need a certificate from an organization saying that someone completed preparing for the professoriate, because we already write letters of recommendation. I can comment in my letters of recommendation for my students to anybody who wants to know that she or he did an outstanding job helping to teach graduate and undergraduate courses, prepare exams, and so on.

Frankly, Karen Phillips, what you have done is preparing you for the professoriate to the nth degree. If I see something like your experience and demonstrated self-initiative on someone's CV and package, it will undoubtedly stand out when I am reading 100 to 200 files for a job search. In particular, it will stand out because your organization of the C^4 program is exactly the kind of thing that we all have to do as professors. We must articulate a need and take our own initiative to do something. Nobody is going to hand it to us. In our institutions, let's facilitate and mentor student initiative, rather than spending great effort in designing new institutional curricular programs. Institutionalization of good programs can be okay, but it is not going to give you the same bang for the buck as mentoring.

Robert L. Lichter, The Camille & Henry Dreyfus Foundation: I want to follow up on that. I have been aware of the Preparing Future Faculty program for some time, and I think the notions are marvelous. I suspect that in many cases it does more, or at least as much, for existing faculty as for future faculty by sensitizing the former to the themes that have been presented here, as Jim Martin just described. I would, however, caution against the notion that institutionalization of these programs is always desirable in all institutions, and that working 70 to 80 hours a week is always bad. Life requires a lot of effort, choices have to be made, and priorities have to be set. You have to decide what you want to do, what you need in order to do it, and then go after what you need.

Members of the faculty are not the graduate students' enemies. To the extent that the issues discussed here are important, and they are, I encourage students to seek out allies among the faculty. They do exist. If Jim Martin says, "my door is open and, yes, I am busy," but he also said to grab him by the neck, then grab him by the neck. And grab some of the others by the neck, too. Most will respond. More faculty are available for you than you realize. Some won't be. Some of those you won't ever have to deal with, but some you will. You have to learn how to do that; that's part of negotiating your way through life. The guidance you can and should get won't necessarily come via an institutionalized program.

I am impressed with the array of students here because of their different career objectives. I hope that everyone here is engaged by and listening to what they are saying. I have one question for the first speaker, Ms. Phillips. I found your array of panelists for that forum very strange—not the people themselves, just the selection. You had two people from within Columbia, a third from an affiliated university, and the editor of *Chemical & Engineering News*, for example. I was curious about why you picked those people.

Karen E.S. Phillips: With our limited awareness, we depended a lot on the advice of Professors Fine and Breslow. We were given a list from which to choose. We were looking for people we thought would present interesting ideas but from different perspectives. We thought Dr. Osteryoung would be able to address the issue of funding for Ph.D. programs. Madeleine Jacobs had interesting statistics about the job market. Ed Wasserman might present ideas that demonstrated what change needed to take place, since running for the ACS presidency might bring him to conclusions and ideas about what needs to be changed. Sally Chapman presented a view of a chair from a small undergraduate university and the pressures of being affiliated with Columbia, but a liberal arts college nonetheless. We had a lot of interactions with Eduardo Macagno as students, and, as the dean of the graduate school, he seemed to be promoting the idea of broadening the curriculum and of encouraging students to do things outside the classic program. He has set up a broad-based ethics program at the university for graduate students, and so on. So, he seemed to have set ideas about making things more interdisciplinary, and I know he had been working with the department for quite awhile trying to get professors to be more cooperative. From our point of view, we were interested in a broad base of ideas, and we thought that these people would best serve the issues. We wanted to address issues such as the value of the Ph.D., where the Ph.D. was going, and what things might affect how the Ph.D. might change or not change.

Victor Vandell, Louisiana State University: I would first like to commend all the panelists for what they said. Some very important issues were brought out about various obstacles faced by graduate students. One thing I want to call attention to is that it is obvious that graduate school is hard work. I could hear people chuckling at all of the things that were painful in the graduate school experience. I guess it is easy to laugh when you are on the other side of the fence. I have talked to a lot of graduate students and they share the sentiment that they do not regret choosing this path and—similar to the theory of no pain, no gain—it will all be worth it in the end. I still believe that graduate school is somewhat exclusive in that you have to be lucky enough to have a mentor who influences you early in your academic life and steers you in that direction. A lot of students don't get to graduate school because they aren't lucky enough to have had someone pull them aside and ask them if they had ever thought about graduate school or to tell them what graduate school is about.

I have been working to bring down the wall that stops students from continuing in science. I tell as many undergraduates as I can that no matter how much hard work this is, you won't regret it, and you will only benefit from it. My question to the panelists is, What are you doing to bring your experience to others and influence their lives? In other words, what are you doing to become role models that other students can follow down that path? Would you suggest other students follow the path you have taken?

Judson L. Haynes III: I can address this in two parts. One activity that I did in graduate school was to go out and give chemical demonstrations. This allowed me to get back in touch with the real world after being required to spend long hours in the lab. It gave me an opportunity to reaffirm to myself that what I was doing was important, could affect people's lives, and could excite the youth. So, I think that that is one mechanism of giving back. Second, now as a professional, I give back by tutoring kids, and I think that is important. When I was in high school, I didn't have anybody that I could go to for tutoring in chemistry. What I try to do is set up or establish something so that students who have the desire or the initiative and want to learn can come to me and I can tutor them. If someone is interested in chemistry and they want to find out who can help them, they can come to me.

Karen E.S. Phillips: Let me add to that. Like Judson, I came through the Minority Access to Research Careers program. Every year, they have a minority research program at Columbia, and every year since I have been there I go and talk to the students at the beginning of the summer session. I tell them what it is like to be in graduate school and what they can hope for. At the end of the summer, I help them with preparation for their final presentations and so on.

Edwin A. Chandross, Bell Laboratories, Lucent Technologies: I want to draw attention to what I consider a critically important thing that, except for a brief mention by Karen Phillips, has been overlooked—the interview process. I have a fair amount of experience interviewing people. I did on-campus interviewing at two institutions in Cambridge, Massachusetts, for many years. I have probably done more than 500 interviews. Let me tell you, you need to help your students with it. They really are unprepared. Let me punctuate this with a brief story of a personal experience. Forty-one years ago, I turned up at Bell Labs and was asked a question for which I was totally unprepared. I think the answer got me a job at a rather demanding institution. What is the value of your Ph.D. research? Being quite young—I had just turned 24—and naive and honest, I told them the truth: there is no value in it besides getting me a degree. I would never do anything like that again. But seriously, I think that this is something that the faculty needs to spend time on with the students and help them. You can give them a leg up by having them be better prepared.

Edel Wasserman, E.I. du Pont de Nemours & Company: I would like to pick up on Dr. Haynes's comment. The ACS is everywhere and can be very useful. There are 188 local sections, and student affiliate chapters exist at many colleges. These affiliates are groups that could be of use for some of the activities that Richard Weibl was addressing. It is easier to be a representative of a formal organization, when asking a faculty member to address a group, than to be an unassociated individual. The ACS has a great deal of support available within its Washington, D.C., headquarters. If you have a question on how to do something, there is almost always someone there who can provide an answer for you.

With respect to Ed Chandross's comments, we are trying to start communication workshops, which are focused on improving the skills people need in interviews on campus. There is a good deal for students to learn there. Communicating thesis work in a largely nontechnical way to trained scientists outside one's own field, avoiding the jargon, and yet communicating the substance can assist the interviewing process. This skill is also necessary throughout one's career as we deal with others of varied backgrounds.

12

Broadening the Scientific Ph.D.: The Princeton Experience

François M.M. Morel
Princeton University

It is a banal observation and a source of oft-repeated jokes that academic specialization is getting out of control: we all seem to know more and more about less and less and worry that soon we might know everything about nothing. At the doctoral level, graduate education is designed to reproduce ourselves, so that we expect our graduating students to know more about some arcane bit of our own field than we ourselves do and than anybody else in the world does. We value nothing more than complete dedication to research and scholarship. We are suspicious of students with outside interests and activities and worry that they are not "serious enough" about science to succeed in research or academia.

The reasons we organize our educational system this way are obvious: it works, and it is self-replicating. Our understanding of the world is slowly increased by all those doctoral students who are pushing, with sharp tools and pointed questions, small corners of the envelope of human knowledge. These students also make a name for themselves in the process. They can then have successful research and teaching careers in which they train the next generation of students to do the same.

The problems with this system are also obvious. Not all doctoral students pursue a research career or stay in it for their entire lives, even in academia. They are then poorly prepared to deal with the broader, nonscientific questions that are part of their jobs; in some important way we fail these students in their doctoral education. We also fail society, which needs to have trained scientists participating effectively in the management of public affairs and private concerns.

Mindful of these issues, an experiment in broadening the doctoral education of scientists and engineers was undertaken in spring 1997 at Princeton University under the leadership of Robert Socolow of the Mechanical and Aerospace Engineering Department, and Thomas Spiro of the Chemistry Department. The program, first dubbed PEI-RISE (Princeton Environmental Institute-Research Initiative in Science and Engineering), and hereafter, PEI-STEP (Science, Technology, and Environmental Policy) has received support from the Chemistry Division of the National Science Foundation (NSF) and is administered by PEI. The program is addressed to all science and engineering Ph.D. students with interest in the environment and aims at providing them with a solid foundation in environmental policy without decreasing or diluting their technical or scientific training. To make the program an exemplar of what a broadened doctoral education could be, we also aim at attracting the strongest possible students, and we organized PEI-STEP as a competitive fellowship program.

To ensure that the environmental policy aspect of their education is substantive, PEI-STEP students must take a series of three courses (two required, one elective) in the Woodrow Wilson School of Public and International Affairs and write a policy research paper. The courses qualify the students for the Graduate Certificate in Science, Technology, and Environmental Policy from the Woodrow Wilson School. The policy paper is included as part of the students' theses and intended to be of publishable quality. To help in this endeavor, PEI-STEP students have a second advisor (often from the Wilson School), who supervises their environmental policy research.

The obvious challenges from adding such substantial requirements to a doctoral program are these:

1. How to support the students financially while they are not pursuing funded research or helping teach in their home departments; and
2. How to fulfill the added requirements without unduly lengthening what is usually an already long graduate career (at least by historical standards).

The first problem is solved fairly adequately by providing half-fellowship support to the students for the two years that they are devoting part of their time to policy work. The second problem has no good solution, and we simply hope to mitigate it by selecting particularly strong students. This makes the selection process critically important. In addition to having a good academic record and strong letters of recommendation the PEI-STEP candidates are expected to present a detailed plan for their policy research. This research plan is developed with the help of the PEI-STEP coordinator and the student's would-be policy advisor. (See the 1999 PEI-STEP advertisement in Attachment 1.)

The organization of the PEI-STEP program resulted from thoughtful reflection, careful planning, and numerous discussions. Three years after its inception, how in fact has it worked? What effect has it had on Princeton's graduate education? What have been the consequences for the students involved?

The first conclusion to be drawn from the PEI-STEP experiment is that such a program is of interest to only a small number of students. The potentially interested population, i.e., the graduate students with a professional interest in the environment at Princeton includes a sizable fraction of those in civil and environmental engineering (CEE), geosciences (GEO), and ecology and evolutionary biology (EEB); a smaller, unknown fraction of those in chemistry (CHM), chemical engineering (CHE), and electrical engineering (EE); and a smattering of individuals from other departments. The pool of possible applicants is probably on the order of 20 to 30 per year, somewhat more than 10 percent of the graduate student cohort at Princeton. The program has gotten 22 completed applications over three rounds and only a few more expressions of interest; a total corresponding to about a third of the potential pool. This is so even though the environmental field is unusually well suited for a broadening of the Ph.D. education: the students it attracts have interests that go beyond pure science and in their careers they will often benefit from a background in policy. Thus, a generalization of a PEI-STEP-type program to other fields would likely attract an even smaller proportion of eligible students.

The 14 students who have enrolled in PEI-STEP over the past two years are distributed evenly among five departments (CHM, GEO, EEB, EE, CEE) and between engineers and scientists. Their projects have ranged widely, encompassing analysis of environmental risk, economic cost, and ecological impact (see list of publications, Attachment 2). The quality of these projects has also ranged widely from modest papers to full peer-reviewed articles. As seen in Attachment 2, of the first 10 projects, at least 5 should result in published articles. While the numbers are too small at this point for useful statistics, it does not appear that the PEI-STEP program has much lengthened the graduation time of the enrolled students—perhaps 3 to 6 months based on examination of individual cases.

The impact of the program on the long-term career of the individual PEI-STEP students is likely to be profound. At least one student in each of the first two cohorts is likely to pursue an academic career at the interface of science and policy. These two students have effectively changed their career plans as a result of their enrollment in the PEI-STEP program. Of the other students, several—perhaps half—will pursue careers in public service where their policy background should prove very helpful. These students, like many with interest in the environmental field, want to make a difference—they have found their foray into policy to be enabling.

Despite the relatively small number of students enrolled, the PEI-STEP program has had an impact that goes beyond the careers of the individual students. A surprising finding in our review of the program was the general opinion, among the regular science advisors, that their whole research group had been affected by, and benefited from, the experience of their PEI-STEP students. Broader discussions had resulted, and a number of students not enrolled in PEI-STEP had decided to take policy courses.

The program has also had an impact on the recruitment of new graduate students: at welcome parties for new graduate students in various departments, a number of students explained that they had been attracted to Princeton by the existence of the PEI-STEP program. Interestingly, there is no evidence at this point for a resulting increase in the PEI-STEP applicants. In fact, several students who had explicitly stated their initial interest in the program have not applied to it. It seems that they are now immersed in the prevalent ethos of the graduate school: they think that enrolling in the PEI-STEP program would somehow mean that they are "not serious" about their scientific work—or, more crucially, they believe that it might appear that way to their advisors.

Clearly, this last observation is the key to the issue of broadening the doctoral education of scientists and engineers, and it may serve as a useful conclusion to this paper. The personal attitude of the individual advisors effectively controls the participation of graduate students in any such program. In many cases, as a matter of course, advisors will simply not allow their students to apply. In others, the explicit or implicit message is that such programs are for second-rate students who are not able to pursue a career in research or academia. Practically all the students enrolled in the PEI-STEP program at Princeton have come from research groups where the professor takes an unusually wide view of graduate education. These professors are themselves interested, and often involved, in issues wider than strictly scientific. (No doubt this is a major reason for the positive assessment we have gotten in our reviews of the program.) As a result, we should expect that programs designed to broaden the doctoral education of scientists and engineers, though perhaps very important, will only attract relatively small numbers of students. Any attempt at generalizing such programs will knock against the hard reality of the self-perpetrating ethos of narrow specialization of academic scientists.

ATTACHMENT 1

ENVIRONMENTAL POLICY FELLOWSHIPS FOR PH.D. STUDENTS IN SCIENCE AND ENGINEERING

❑ **The Princeton Environmental Institute** announces the fourth annual competition for the Princeton Environmental Institute Science, Technology and Environmental Policy (PEI-STEP) Fellowships. Half-time fellowships (stipend and tuition) are provided for 24 months, to permit Ph.D. students in science and engineering to address environmental policy implications of their thesis research through supplementary course work and policy-oriented research.

❑ **The goal of PEI-STEP** is to make students more effective and more versatile in their careers as scientists, teachers, and leaders in the public and private sectors and to increase awareness among science and engineering students and faculty of how their discipline-based skills can be brought to bear on environmental problems.

❑ **PEI-STEP Fellows** have an additional advisor from another department who, in cooperation with the primary advisor, will supervise the environmental policy research. The student writes an independent paper on the policy research, the equivalent of a chapter of the student's thesis. In addition, PEI-STEP students are awarded the Graduate Certificate in Science, Technology and Environmental Policy from the Woodrow Wilson School of Public and International Affairs. To meet the requirements of this certificate, the student normally takes three courses in aspects of science policy and technology assessment.

APPLICATIONS:
Currently enrolled graduate students in their first, second, or third year in all science and engineering departments are eligible to apply. The application should include a c.v. and a detailed research plan, worked out in cooperation with the student's thesis advisor and the proposed PEI-STEP advisor. Letters of support from both advisors are required. Criteria for selection includes a strong academic record, a well-thought-out research plan, and engagement of the thesis advisor in the research plan. PEI-STEP fellowship support will begin in the summer or fall of 2000.

Application forms and additional information are available at <www.princeton.edu/~pei> or at the PEI office, Guyot 25. Prospective applicants should contact Valerie Thomas (address below) no later than April 3, 2000. Final applications are due April 30, 2000, at the PEI office, Guyot 25.

Valerie Thomas
Center for Energy and Environmental Studies, H-214 E-Quad.
Tel: 258-4665. E-mail: vmthomas@princeton.edu.

ATTACHMENT 2

PAPERS (PUBLISHED AND ADVANCE DRAFTS)

"Risk Assessment for Polycyclic Aromatic Hydrocarbon NAPLs Using a Fraction Approach." D.G. Brown, C.D. Knightes, C.A. Peters. *Environmental Science and Technology,* December 15, 1999.

"The North Atlantic Thermohaline Circulation Collapse as a Constraint on Economic Optimal Carbon Dioxide Emissions." K. Keller, K. Tan, D.F. Bradford. 1998 Draft.

"Commercialization of Photovoltaics: Long-Run Cost Estimates." A. Payne, R. Duke, R. Williams. September 1998 Draft.

"Worker Exposure and Health Risks from Volatile Organic Compounds Utilized in the Paint Manufacturing Industry of Kenya." K.L. Purvis, I.O. Jumba, S. Wandiga, J. Zhang, and D.M. Kammen. Submitted to *Environmental Science and Technology.*

"Ecological Impact of the Venezuelan Economic Crisis." J.P. Rodriguez. August 1999 Draft.

13

Across the Disciplines: Center-based Graduate Education and Research

J. Michael White
University of Texas at Austin

Chemistry has been called the "central science" and, within the energy regime thermally accessible to anything dealing with molecules and materials, I suppose it is so among the natural sciences. Assuming this is the case, one can then argue that chemistry can also be called the "decentralized science"—decentralized in the context of those traditions we all know about that name, separate, and distinguish among groups of people, organizations, departments, and the like.

For example, my own research area, the study of surfaces and interfaces, has had an "up and down" popularity within the tribe of physical chemists over the last 100 years. After Langmuir, the tribe of surface chemists became nearly extinct, but was then resurrected in the 1960s. I still remember going through the American Chemical Society (ACS) Directory of Graduate Research in the mid-1960s looking for chemists focusing their research on surface and interface science. I counted them with the fingers of one hand. And I remember arguing for incorporation of the name "surface chemistry" into the program names within the chemistry division of the National Science Foundation (NSF) in the late 1970s. It takes time for these cultures to change. From my point of view, our tribe needs to be at the forefront in every conceivable way—organizing and reorganizing ourselves on a periodic basis to avoid becoming a hopelessly arthritic culture. The 11-year life, with an enforced sunset, of the interdisciplinary NSF Science and Technology Centers (STCs) program[1] is a reasonable timescale for reinvention and reorganization.

In my own case, moving from gas-phase reactions and dynamics to surface reactions and dynamics required a modest bucking of culture in the late 1960s. Today when I look at promising young Ph.D. graduates, two questions I always ask are: "Could they reinvent themselves and could they do it in two years? and "Would they look forward to doing it?" Many of my colleagues in the microelectronics industry tell me they need to reinvent themselves at the master's degree level every 18 months, just like

[1]Information on the Science and Technology Centers program, other interdisciplinary programs at NSF, and links to other interdisciplinary funding sources can be found on the World Wide Web at <www.nsf.gov> under, among others, "Cross-cutting Programs" and "Office of Integrative Activities."

Moore's law for chips, if they are to maintain a productive and up-to-date industrial career. It is my conviction that such an attitude needs to be instilled in our graduate education enterprise: "Look forward to reinventing yourself at least every decade." I have seen that anticipation among the young faculty at the University of Texas at Austin, and I see it among the young faculty at this meeting. Let me say to you young faculty members, press on and be very inclusive engaging your colleagues from other sciences, business, and humanities—and plan to reinvent yourselves every 5 to 10 years. For the older ones among us, they say, "You can't teach an old dog new tricks," but I think it can be done, provided we starve the old dog for a time.

For about a decade, I have directed an NSF STC dealing with the synthesis, growth, and analysis of electronic materials, an 11-year interdisciplinary program that has integrated graduate research, and the education that attends it, through collaborations between faculty in electrical engineering, chemistry, physics, and chemical engineering.

Why a center? For one reason, as in many other areas of science, engineering, and technology, there is a host of interesting questions that are not resolvable or even addressable by the traditional isolated individual-investigator mode of attack that has traditionally characterized my own work and most of my tribe. What we need for complex problems is a group of strong, capable individual investigators, i.e., those quite capable of operating successfully on their own, who come coherently together focused for a time with resources sufficient to attack complex problems—many of which centrally and intrinsically involve molecular-level chemistry coupled with all those complexities that link them to integrated systems—something like the words of the prophetic vision in the "valley of the dry bones." But rather than thigh bones and leg bones being linked, we might say that the atoms connect to the molecules that connect to the nanoscale structures that connect to the mesoscale structures that connect the devices and cells that we make and use—in other words, the fundamental chemistry of integrated systems. In my view it is the nature of the problems that should drive interdisciplinary center-based research. Funding center-based research for a decade timescale, as in the NSF STCs program, focuses for a sufficiently long period of time the necessary faculty, facilities, funds, and students—something highly unlikely for single-investigator support. Developing central multiuser facilities is one critical benefit of center-based funding. Regarding naturally multidisciplinary research problems, there are, for graduate education in chemistry, wonderful opportunities and challenges in materials sciences, optical sciences, health sciences, neurosciences, and environmental sciences. Chemistry would do well to define itself for the purposes of graduate research education "in" rather than "out" of these arenas.

In the case of our center, we have targeted a number of fundamental issues regarding how the properties of molecules (precursors) are related to the properties of electronic material thin films grown from them, how to make novel multi-element films with desired electronic and optoelectronic properties, how to control and analyze films as they grow, and how to link chemical properties with electrical properties of interfaces. Our progress has relied on many long-term collaborations among faculty, students, and postdocs from chemistry, chemical engineering, physics, and electrical engineering. Collaborations that, once initiated through the STC, have developed complementary funding that will live on and evolve well past the sunset of the STC.

Again, why a center? For another reason, the professional lives of most of our graduates will involve defining, addressing, and assessing often complex issues that benefit enormously from multidisciplinary education and multidisciplinary experience. While course work is of some value, graduate training in this direction is, in my view, best accomplished when graduate students, well trained through fundamental course work in each of their various disciplines, work productively alongside each other and communicate daily. We have realized this kind of "hands-on" education with many of the

25 students supported directly by our STC. We do this by engaging them in interdisciplinary research that forms the bulk of their Ph.D. dissertations. At a lower level, but one that has had considerable benefit, our in-house student-based scientific meetings twice each year have provided a regular forum for exchanging ideas and information at the graduate student level. From this perspective, center-based research programs can be splendid vehicles for preparing graduate students for dealing capably with the broad-based, problem-solving situations they are likely to find in their professional lives.

What is the response of the students and potential employers? As usual, the response is dependent on the quality of the students, but it is clear, for example, that our students who have worked jointly between chemistry and electrical engineering are uniquely qualified to engage in chemical vapor deposition (CVD) process research and development at major microelectronics firms, and they are doing so successfully. This is particularly true for our STC postdoctoral colleagues, each of whom has a research job description involving interdisciplinary accountability. Not only are these graduate students and postdocs uniquely qualified, they are uniquely attractive to commercial firms. This comes as one result of their having dealt in some depth with both molecular-level chemistry issues and device-performance issues. As a group leader from one company recently put it: "This Ph.D. student is uniquely attractive to my company because she is comfortable with both the chemistry and the electronics aspects of CVD processing and can walk right into our work environment and contribute immediately."

For how long will such an education be of great value? To the extent that it encourages individuals to think in terms of life-long education in new areas (reinventing oneself), it lasts for a lifetime.

What are the barriers to center-based graduate education? Perhaps the question can be changed to, What are the issues? As in all cultures, they exist and, in my opinion, should and will always exist. These tensions and struggles can actually be of great benefit. Academic deans, charged with representing their faculty, have college funding and course work concerns. Department chairs rightly have concerns regarding faculty loyalties. Faculty members have concerns regarding independence and promotion. Center directors have concerns regarding sunset issues with staff and central facilities while striving to develop effective, flexible, and evolving infrastructure.

Since most discussions about the future of graduate education and research in chemistry presume a supply of students, the following are worthwhile questions, in my opinion:

- In 10 years, who will comprise our graduate student population?
- Compared with their other options, why would U.S. undergraduates want to go to graduate school in a chemistry department?
- In the United States, are not our graduate chemistry departments competing over a shrinking pool?
- If we were to set out to make sure that the number of students interested in entering graduate work in chemistry met the needs and demands of our country, what tasks would we undertake?

The last question offers one opportunity for considering how to expand the pool of talent and, since it is a key part of our STC work, I address it here. What would we do to make sure that a decade from now that adequate numbers of students enter graduate work in chemistry? Among other things, we would take initiatives to open our universities to our communities in nontoken, long-term, sustainable ways that bring, for example, sixth graders, especially those from communities of poverty, through the educational system, helping them maintain a serious interest in undergraduate education. We would see them develop the skills to compete and make productive contributions to our society. In other words we would undertake a sustainable, long-term, steady, and low-level shepherding role, from sixth grade to undergraduate degree, for young men and women from our communities of poverty.

That is one precollege educational goal that our science and technology center takes seriously and has included as part of our charter. Our Young Scientists Program has enjoyed nine years of success, starting in a single school and now involving eight elementary schools. As one measure of its success, this program is being taken over cooperatively by the University of Texas and the Austin Independent School District. Significant numbers of students from elementary schools in generally financially poor areas of Austin are being assisted in responsible ways to succeed in academic life.

What is the Young Scientists Program? It is a hands-on, sixth-grade, student-focused, and science-based classroom program designed to encourage and assist children, especially those from communities of poverty, to enter high school and, later on, college, with excellent academic credentials. Young Scientists involves the students, their teachers, principals, and parents, and personnel from the University of Texas in a sustainable program focused, where it must be, on the students.

Related to this, I have just completed lecturing to two huge sections—roughly 450 students in each lecture—of general chemistry for science and engineering students. Not surprisingly, I am both encouraged and discouraged. All of these students are quite familiar with the huge libraries of information at their fingertips, or should I say their computer screens, but I am constantly appalled by their generally horrible experience with simple mathematics. Here is an example: "If $K_a = 10^{-6}$, what is pK_a?" It turns out that doing this on one of the more popular calculators, in what appears to be an appropriate entry method, yields "5" for the answer. You do not want to know how many second-semester chemistry students answered "5" and quarreled with me "because this answer came out of this machine." This issue makes me keenly aware of ground often lost between sixth grade and the end of high school—ground that is difficult, I might say nearly impossible, to recover upon entering college. My conclusion is that high-quality education must be, by and large, a linked seamless process, and weak links anywhere in the chain are disastrous.

I am of the opinion that investments at the sixth grade and forward through high school are critical to the future of graduate education in chemistry. We do not need novel solutions. We need to follow the well-worn advertising line, "Just do it." We need to do what we know works—and what works is individual, personal investment of our time and, yes, our personal money, over a decade to influence a few students and see them successfully through. The rewards will be magnificent even as we adapt to the failures, resistances, and rejections that inevitably come. I think we know what has worked and what will work. The question really is, What are we willing to invest of our individual resources?

To close, let me just observe how center-based graduate education in the STC framework has influenced my own work. More than 10 years ago, I became interested in chemical vapor deposition and thin film growth from molecular precursors but realized that the benefits would be enhanced were I to learn the electrical engineering that enters into device considerations. That has been realized through long-term, center-based collaborations with two colleagues, one in chemical engineering, the other in electrical engineering. We have made excellent use of central facilities made possible by the focused long-term funding undergirding our STC. We have jointly supervised a number of graduate students and, along the way, I have learned the electrical engineering concepts and benefited enormously from the intellectual stimulation. And I have the clear sense that students doing research in such a day-in, day-out environment graduate looking forward to tackling complex problems and reinventing themselves every 5 to 10 years. Our country needs them and more of them.

DISCUSSION

Jeanne Pemberton, University of Arizona: I would like to address one aspect of the graduate experience that you commented on, and that is the use of formal course work. You said that you couldn't really teach graduate students what they need to know in courses. I would like to take exception to that.

Perhaps by teaching them in the traditional sense, the so-called telling mode as Angie Stacy called it yesterday, we are not doing our students a great deal of service. Perhaps we could be more creative in the kinds of things that we do in our courses to educate our students broadly. I am referring to things that are more experiential than lecture based. Could you comment on activities that you perhaps use in your center that are experiential and give students a broader range of skills?

J. Michael White: We have twice annually what I will call a miniature scientific meeting. That is one thing. Second, I really don't want to get too extreme about the use of course work. We have a team-taught interdisciplinary course available not only to the science and technology center student, but to all students who deal with the subject matters that are the focus of the center. There are also many one- or two-day tutorials on various aspects of instrumentation. These are activities that the science and technology center does three or four times a year. I don't have the numbers in front of me.

Jeanne Pemberton: Do you do anything that is laboratory based for graduate students? I know that is heresy to suggest, but I think that there are ways we can broaden the skill set of graduate students by implementing laboratory-based kinds of activities that are short term but not actually research.

J. Michael White: There are two cases that I can think of. There is a short course on surface analysis and a short course on electron microscopy that is a component of our center's activity. Those are given once a year.

John Schwab, National Institute of General Medical Sciences: My grant portfolio includes organic synthesis, medicinal chemistry, bioorganic chemistry, and natural products chemistry. My question might be more appropriate for Ron Borchardt, but I would like both of you to address it in turn. I find the science and technology centers to be very exciting, because of the potential for conducting highly integrated, interdisciplinary scientific research. However, I am concerned about the job prospects of students who are trained in such an environment. With few exceptions, when hiring medical chemists, the pharmaceutical industry focuses its recruiting efforts solely on students who have a great deal of depth in organic synthesis but not much scientific breadth. Particularly in the context of organic synthesis, it is perceived that breadth of knowledge comes only at the expense of depth of knowledge. I wonder if you have concerns about the marketability of the students who are trained in highly integrated, multi-disciplinary science?

J. Michael White: I would go back to what I intended to say, and that is that I don't think we have compromised the depth at all. We have moved the drilling bit over onto the interface between what has traditionally been called chemistry and what has traditionally been called electrical engineering. I had a call a week ago from a staff member of a major semiconductor company. He said, "I have never seen a student like Allen Mao. He can talk the language of molecular chemistry, and he can also talk the language of device engineering." Allen is an electrical engineer. He has a Ph.D. This is anecdotal because we don't have huge numbers. I can give you the numbers, but they won't mean much to you. There is no doubt that you can educate too broadly. That is not my goal. My goal is to move the drilling bit over and to get a look at both sides in a different kind of way if you want to establish a new kind of discipline.

John Schwab: Of course, when you move the drilling bit over, the product that you are getting out is somewhat different. Then again, assuming the product is deeply trained, it is a question of whether or not the product is still going to be marketable. Perhaps this is a difference between surface chemistry and electronics, for instance, and pharmaceuticals.

Ronald T. Borchardt, University of Kansas: I am going to address part of this in my talk, but let me add one thought. One of the difficulties in giving students more breadth in their educational experience is the reluctance of the faculty to look critically at the courses that they are teaching and objectively say that in order to add to this list of offered courses we have to delete something. Faculty tend to be hung up on teaching a certain course their entire career, and they are very reluctant to release that course. You have to think about what is in the best interest of the students and their careers. You can have depth in the program and breadth also. But it requires serious discussions and compromise on the part of faculty in departments and in interdisciplinary programs.

C. Dale Poulter, University of Utah: John and others brought up issues related to the importance of attracting people into the profession from diverse backgrounds and areas of interest because of the shrinking pipeline of students. From the discussion, the participants in this workshop recognize this problem. I think there is another aspect related to diversity that is more subtle but also very important to the profession. The selection of specific research problems can often, I think, be influenced by one's background. An example that I would cite is the field of ethnobotany, where one uses insights obtained from folk medicine to search for new pharmaceutical agents, typically from native plants. An Hispanic-American colleague of mine became interested in the Aztec language and from that interest had occasion to view remnants of Aztec writing related to the sophisticated folk medicine that they had developed before the Spanish conquest. His interest in Aztec history provided him with the background to identify specific plants and their related medicinal use. He then continued by identifying specific chemical compounds in the plants that were responsible for the pharmacological effects. This work was initiated more than 25 years ago. It was one of the first uses of folklore to provide leads for drugs and just one example about how one's culture and background can influence the direction of research. I think it is something we don't consider.

R. Stephen Berry, University of Chicago: We know, of course, that there is a wide range of styles from one university to another. I am curious about how much you think you could have accomplished in this integration of electrical engineering and surface science in the absence of a formal structure. If you had gone informally to people in electrical engineering and said, "let's try to do this together," how far do you think you could have gone?

J. Michael White: I have thought about that question quite a bit, and I think we could have made some progress. The issue is getting sufficient resources in the same time frame. There are also infrastructure issues. We have central facilities that we have built up and maintained that provide a lot of the longer-term glue that establishes a tradition, if I can put it that way, of continuing our collaborations from one problem to another. The reason I am saying we could have gotten part of the way is that I see my colleagues putting together programs based on one graduate student at a time. But, if properly managed, a kind of coherence arises out of building on a large, more focused base.

R. Stephen Berry: Let me ask about the other side, using examples from my own institution of the Enrico Fermi Institute and the James Franck Institute as models. The institutes were not really set up to address anything as specific as your STC, but have essentially for 50 years been like centers in that they have brought together people from different fields with the problems that they were evolving through that entire period. I think it would be very difficult, or at least unconventional, to try to set up something like that today in contrast to 1946. Would you say anything about how that kind of a structure seems to you, especially since you are coming to the end of your 11 years.

J. Michael White: I would hope that the Texas Materials Institute in 50 years will have done exactly what you are talking about, that it will provide a longer-range, broader vehicle, depending on how it is operated, and managed, for people going in and out of it on roughly the timescale of a graduate student's career and solving and making progress on certain specific problems in that time frame. I hope that happens, and I think we have that infrastructure within the University of Texas at Austin now to realize that goal.

Iwao Ojima, State University of New York at Stony Brook: I came here to listen to all the educational ideas, but I am also very interested in how chemistry goes into the 21st century, as that is the title of this workshop. I firmly believe that interdisciplinarity is a very big key for success, and then maybe chemistry will go into materials science and the rest of science. Then for the materials science side, you will have quite a success with operating your center.

At Stony Brook we have many different incentives. We have two state-operated centers for advanced technologies: one is the Center for Biotechnology, and the other deals with sensors. The centers have been quite successful. We have also started centers for molecular medicine, which includes immunology, genetics, and structural biology. In all of these centers, chemistry is involved. I am very curious about your science and technology center. Is your operation a facility that is shared, or do you appoint faculty members to the center?

J. Michael White: We have 12 faculty members, whose composition has changed slowly over the 10 years of the program. They are not appointed in terms of faculty appointments. They belong to specific departments and rise or fall on the basis of what they do in those departments, a piece of which I suppose is what they do in our center.

Iwao Ojima: So, the faculty and the departments have a share of the instrumentation, but there are no faculty members "living" in the central facilities.

J. Michael White: I would never put a faculty member in that position if I could help it.

Iwao Ojima: We have experimented with one type of arrangement in our structural biology center. Although we do not have core members, a certain number of faculty members have to reside in that center. In addition, other members may be associated. In each case, the faculty member has his or her own home department. This setup is difficult for the junior appointment.

J. Michael White: I can guarantee you that you will have some difficulties in operating it.

Iwao Ojima: Yes. Junior faculty gain by interacting with other faculty members from different departments at the center. Unfortunately, they may lose some contact with their home department. Then, when the time comes for a tenure decision, they suffer from not having spent enough time establishing relationships within the department. Does your system work in a similar style?

J. Michael White: Absolutely not.

Iwao Ojima: Then there is some faculty difference between the traditional chemistry department and biology. Graduate students are recruited into this program but still have a home department. This is sometimes very beneficial, for example, as a training ground for the chemistry/biology areas. Do you think that this type of setup would be effective for chemistry students in the materials science area? Also, the biology-related program has rotations. Do you think rotations would work in this area?

J. Michael White: It is possible. I think it would be administratively frustrating.

Iwao Ojima: I agree, but for students it is quite an experience.

J. Michael White: Yes, it is quite an experience, but I would be careful with respect to what the impact is on their future. Those students in the end need to have a real home.

Participant: I would like Mike to be more specific about the frustrations and about what he thinks about these rotations through departments.

J. Michael White: The students initially apply to a specific department. They have allegiances, and stipends come from specific departments. A certain amount of rotation occurs in terms of faculty presentations, but they don't move from one discipline to another. I accept people who are entering graduate school in chemical engineering, but they would not come into my group on the basis of the formal aspects of the science and technology center. It would be on the basis of a presentation that I would make to the graduate students who were coming into chemical engineering; similarly, in materials science.

We have a materials science and engineering program as an umbrella organization. That is a part of the Texas Materials Institute. It will involve for some people a rotation. It won't be particularly enforced. The students will be asked what areas within materials science they are really interested in and, on a case-by-case basis, they would be rotated, to use that term. So, I don't in principle object to that; you just have to see to it that it gets done and done well. Otherwise, it doesn't add anything that I can see to the future of a graduate student's career.

Robert Lockhead, University of Southern Mississippi: I don't know if you are the best person to answer this, or if Marye Anne Fox or Janet Osteryoung would be better, but it deals with the STC centers, the largest NSF centers, where the focus is usually 12 multidisciplinary faculty members. In Europe, what they have created are megacenters. The megacenters consist of many universities spread across Europe. For example, in water-soluble polymers, they have 11 universities, called Team Luns, Team Bristol, and so on, with each university focused on a single area. By interacting with a center, the students are automatically exposed to many disciplines. The students also rotate through several of the universities on their way to a Ph.D. So, they are exposed to different cultures and to a much better instrumentation base than any one university could afford. Can you comment on the advantage of STC compared to megacenters, or otherwise someone from the NSF if anyone can comment? Is there any move toward megacenters in this country?

J. Michael White: No comment. Janet, did you want to comment?

Janet Osteryoung, National Science Foundation: Yes. The STCs are the program at NSF that is perhaps the most visible, but, of course, there are such things as the Antarctic Research Center and other activities supported by NSF, which are much more costly and involve many more people. It sounds as if the COST program[2] of the European Community is what has just been described. The COST program has as its entire purpose the communication and integration aspects, and the funds go only for that. It is assumed that all of the research involved in these programs is paid for from other sources. So, the concept is quite a bit different from the STCs.

[2]COST is European Cooperation in the field of Scientific and Technical Research: Technical Committee for Chemistry.

14

Training Grants in the Chemical and Biological Sciences

Ronald T. Borchardt
University of Kansas

INTRODUCTION

In the 21st century, we must realize that academic programs, like companies, generate products (i.e., graduates, knowledge, technology) and have multiple clients (e.g., other academic departments, government, industry) that use these products. Academic programs also must continually update their information about their clients and their clients' needs. For example, U.S. corporations in the 1980s and 1990s have undergone dramatic changes in how they conduct business. In addition to introducing new technologies into the processes of discovery, development, manufacturing, marketing, and distribution of products, U.S. corporations now require their employees to perform within highly integrated teams. Participants in these multidisciplinary teams need not only to be technically competent but also to have the interpersonal skills required to function in a team environment. Academic programs, particularly chemistry departments, have been slow to recognize that it is no longer sufficient to train a graduate student to be the world's expert on a narrowly focused topic. Individuals with narrowly focused areas of expertise are often unable to function effectively on a multidisciplinary team. While depth of knowledge and experience in a particular field is still the most important component of good graduate education, students also must acquire scientific breadth and the interpersonal skills necessary to function as a team member.

In the 1980s and 1990s, academic units in the biological and pharmaceutical sciences recognized the need to change and effectively used predoctoral training grants provided by the National Institutes of Health (NIH) to implement this change. Multidisciplinary predoctoral training grants provide an excellent mechanism to bridge the gaps that often exist between academic units (e.g., departments of chemistry and biology and schools of pharmacy and engineering). These training grants also can serve to broaden the educational experiences of the trainees and to refine their interpersonal skills. However, establishing good multidisciplinary training programs requires that the students' interests take precedence over the interests of the faculty and the academic unit. This means that faculty and/or academic leaders (e.g., department chairpersons) must be willing to step forward and think out of the box about graduate education. They must also be willing to be implementers of change in their home institutions.

As an example of programmatic change within an academic unit and of the role of NIH predoctoral training grants in facilitating this change, this article focuses on graduate education in the Department of Pharmaceutical Chemistry at the University of Kansas over the past 30 years (Figures 14.1 through 14.3). In particular, it underscores changes implemented in the mid-1980s to generate Ph.D. scientists who could compete effectively for jobs in the emerging biotechnology industry.

HISTORY OF PHARMACEUTICAL CHEMISTRY GRADUATE EDUCATION AT THE UNIVERSITY OF KANSAS

Pre-Biotechnology Revolution (1967 to 1990)

The Department of Pharmaceutical Chemistry (also called pharmaceutics at some universities) at the University of Kansas was founded in 1967 by the late Professor Takeru Higuchi. Professor Higuchi, who was trained as a physical chemist, entered the field of pharmaceutical chemistry in the late 1940s when he joined the faculty of the School of Pharmacy at the University of Wisconsin. After moving to the University of Kansas, Professor Higuchi built the predoctoral training and research programs in the Department of Pharmaceutical Chemistry around the philosophy that success in understanding drug actions, controlling drug delivery across biological barriers and to drug receptors, developing stable drug formulations, and creating methods for analysis of drug substances required a thorough knowledge of the basic principles of analytical, physical, physical organic, and organic chemistry. Because the pharmaceutical industry at that time was oriented largely toward the discovery and development of small-molecule-based drug candidates, Professor Higuchi's philosophy was well received, and the department built strong relationships with this industry user of its products. These products included Ph.D. scientists as well as knowledge and new technologies focused on small molecules (see Figure 14.1).

In the 1980s, the pharmaceutical industry began to undergo dramatic changes resulting from the biotechnology revolution. The introduction of this new technology made it possible for the first time to produce large quantities of macromolecules (e.g., proteins) as drug candidates. This change began to affect our graduate program in the mid-1980s when some of these macromolecules reached preclinical development. Interestingly, this change in the industry coincided with my transfer from the Department of Biochemistry to the Department of Pharmaceutical Chemistry to replace Professor Higuchi, who had retired as chairperson. While I fully respected and supported Professor Higuchi's educational philosophy, I also recognized the impact that biotechnology was having on the pharmaceutical industry and the need to implement change in our graduate program to accommodate the changes occurring in industry. The need for this change became more evident when biotechnology companies began to hire our Ph.D. graduates to participate in the preclinical development of macromolecule-based drug candidates. We quickly realized (based on feedback from our alumni) that our Ph.D. graduates were not properly trained to perform this development function. During the next five years (from approximately 1985 to 1990), biotechnology companies continued to hire our Ph.D. graduates but were forced to provide on-the-job training related to macromolecules for these new scientists (see Figure 14.2).

As chairperson of the department, I was then confronted with the delicate problem of implementing change in a program that was highly respected both nationally and internationally. I choose the words "delicate problem" because, while our department needed to begin to train Ph.D. scientists who could function effectively as members of development teams for macromolecule-based drug candidates in biotechnology companies, we needed at the same time to continue to produce Ph.D. scientists who could function effectively as members of development teams for small-molecule-based drug candidates in more traditional pharmaceutical companies.

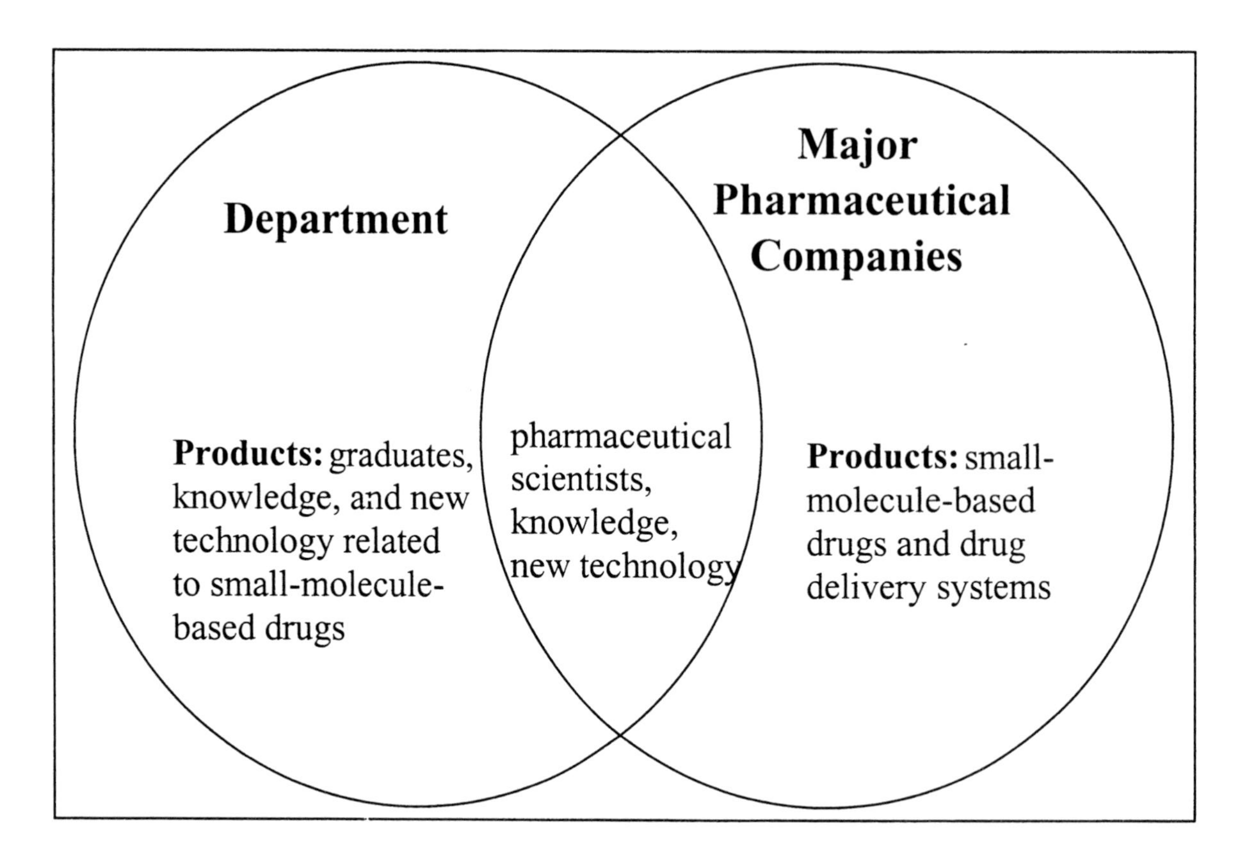

FIGURE 14.1 Pharmaceutical chemistry graduate education at the University of Kansas: 1967 to 1985.

Customarily, academic chairpersons have handled this type of problem by simply hiring additional faculty. While our department was able to add a few new faculty members in the late 1980s, the numbers were not sufficient to produce a dramatic change in our graduate program. The change that was needed involved placing increased research and educational emphasis on molecular biology, biochemistry, biophysical chemistry, and bioengineering as these disciplines apply to pharmaceutical chemistry. This shift was essential because Ph.D. scientists involved in the development of macromolecule-based drug candidates required different training than did Ph.D. scientists involved in the development of small-molecule-based drug candidates (see Figure 14.3).

To gain this training, it became necessary to expand beyond our department's graduate program to create a multidisciplinary training program in pharmaceutical biotechnology. We recruited a subset of the faculty in the Department of Pharmaceutical Chemistry and then carefully selected additional faculty from the Departments of Chemistry, Biomolecular Sciences, and Engineering on our campus (Figure 14.4). In addition, we recruited scientists from biotechnology and major pharmaceutical companies. These scientists were selected because they had expertise in areas not available from our faculty (e.g., bioprocess engineering) and/or because they could serve as mentors for trainees during their industrial internships.

To facilitate the development of this multidisciplinary program in pharmaceutical biotechnology, the University of Kansas successfully competed for a predoctoral training grant from the National Institute of General Medical Sciences (NIGMS) in 1989.

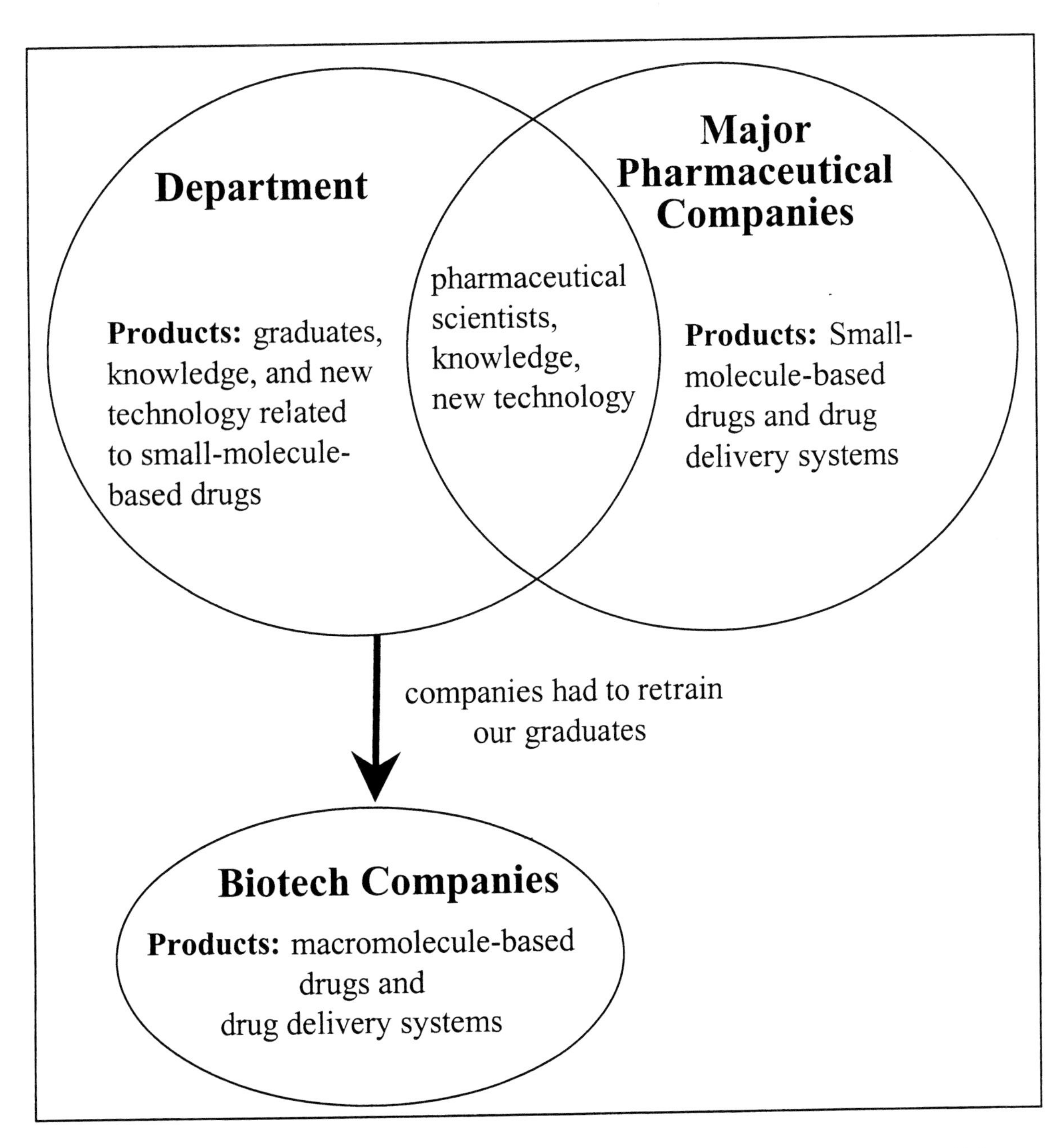

FIGURE 14.2 Pharmaceutical chemistry graduate education at the University of Kansas: 1985 to 1990.

Post-Biotechnology Revolution (1990 to Present)

This multidisciplinary training program in pharmaceutical biotechnology has as its primary objective "the training of pharmaceutical scientists who have the expertise to assist in the preclinical and clinical discovery/development of macromolecule-based drug candidates." The program is administered by a steering committee composed of faculty representatives from the participating academic units (i.e., Departments of Pharmaceutical Chemistry, Chemistry, Molecular Biosciences, and Chemical Engineer-

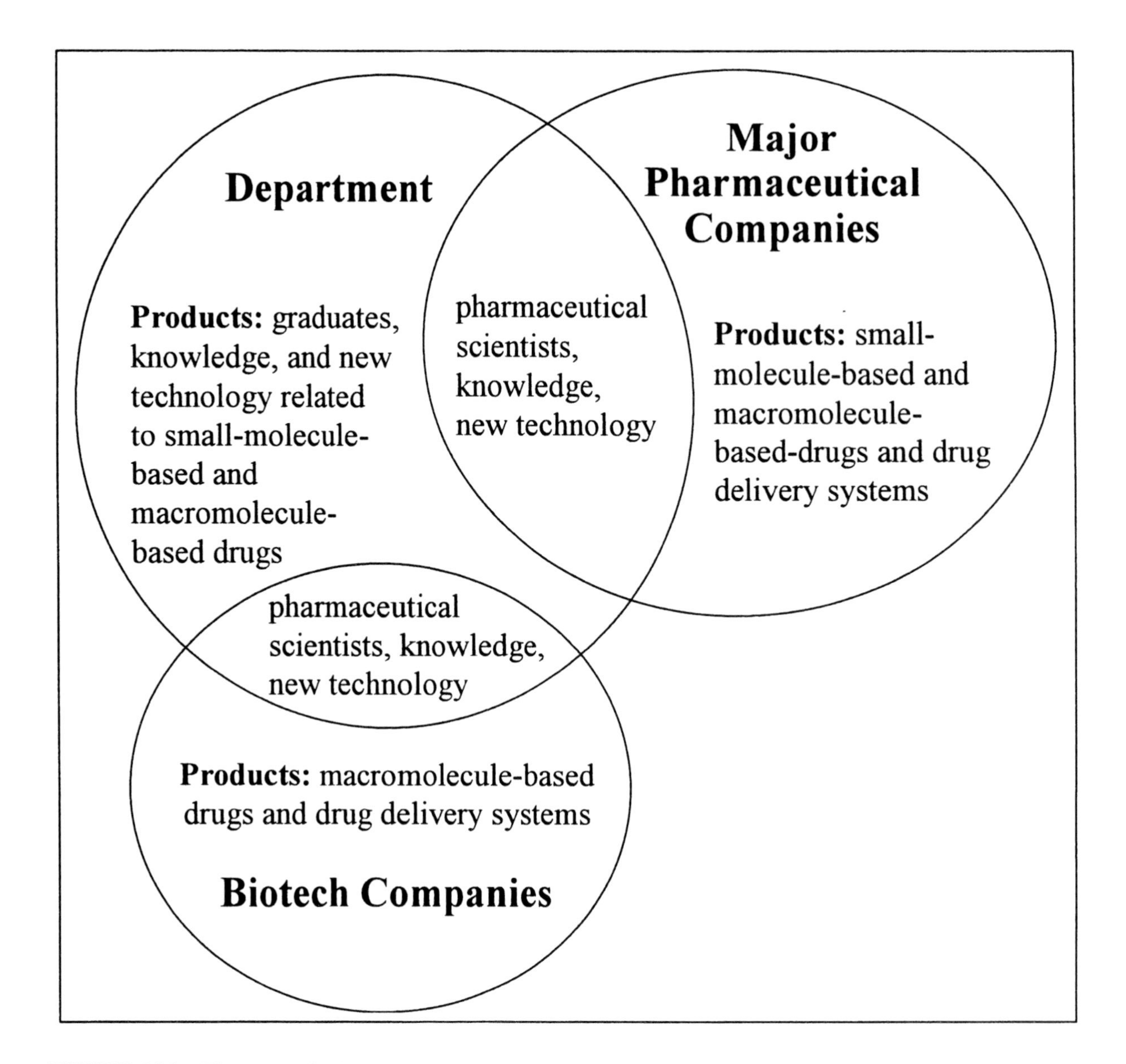

FIGURE 14.3 Pharmaceutical chemistry graduate education at the University of Kansas: 1990 to present.

ing). The trainees for this program are selected from graduate students enrolled in the participating departments who have completed at least one year of graduate education at the University of Kansas. The trainees continue as graduate students in their respective departments. They fulfill the requirements of their "home" graduate program and, ultimately, their Ph.D. degrees are awarded in these disciplines. However, as trainees in the Pharmaceutical Biotechnology Program, the students also must complete several additional requirements, including completion of "core" courses and courses in their area of specialization, participation in a biotechnology seminar and a biotechnology journal club, and completion of an industrial internship. Each of these requirements is described below in detail.

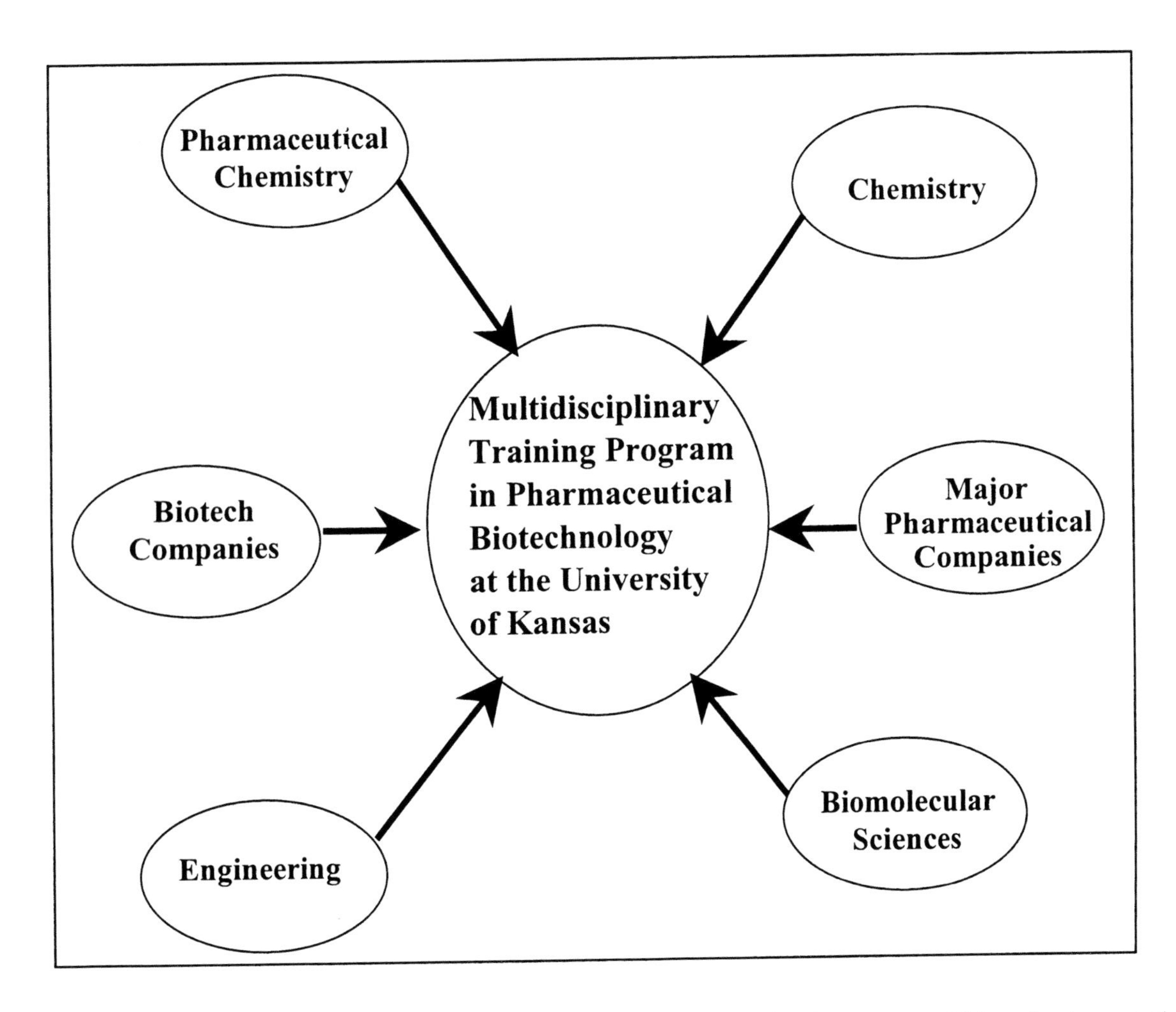

FIGURE 14.4 Participants in the multidisciplinary training program in pharmaceutical biotechnology at the University of Kansas.

Trainees in the Pharmaceutical Biotechnology Program are required to complete three "core" courses, one of which is in molecular biology. From this course, the students learn the language of this discipline, which helps them function effectively in multidisciplinary drug discovery development teams in biotechnology companies. The students also complete a course entitled "Advanced Pharmaceutical Biotechnology," which increases their scientific breadth through exposure to all aspects of the discovery and development of macromolecules as drugs. Topics covered in the course include techniques used to clone and express proteins, scale-up of expression systems for proteins for manufacturing, characterization of the pharmacological and toxicological properties of the proteins, incorporation of proteins into appropriate dosage forms, and regulatory issues related to approval of the drug product by the Food and Drug Administration. Students are also required to complete a seminar course, "Issues in Scientific Integrity," which addresses the multiple roles of scientists as researchers, authors, grantees, reviewers, inventors, employers/employees, teachers/students, and citizens. Trainees also complete two additional courses in their chosen areas of specialization (i.e., delivery, formulation, or analysis). These courses are designed to provide the scientific depth that Ph.D. scientists need to successfully execute their technical role as a member of a multidisciplinary drug development team in a biotechnology company.

To facilitate communication among the trainees and the faculty members and to ensure exposure of the trainees to the external scientific community, the students participate in a monthly seminar series entitled "Pharmaceutical Aspects of Biotechnology." Many of the seminar speakers are scientists from biotechnology and major pharmaceutical companies. These individuals provide the students with practical examples of the development of macromolecules as drug candidates. This seminar series also serves to expose the students to new scientific findings in the broad field of biotechnology.

Further exposure of students to the newer findings in biotechnology is provided through a biotechnology journal club. Each semester, a new topic (e.g., gene delivery, mass spectrometry in biotechnology) is selected, and the students are required to present and discuss papers relevant to this subject. This journal club has helped to provide additional breadth and depth in the students' training.

Finally, the trainees must complete an industrial internship that is at least three months in duration. Typically, students undertake this requirement after they have completed their departmental Ph.D. preliminary examinations and are well into their dissertation research. Ideally, the internship allows them to continue some aspect of their dissertation research in an industrial laboratory or to complete a project that can be incorporated into their dissertation. Students often use the internship to learn new techniques or gain access to equipment that will enhance completion of their dissertation research. The success of this type of industrial internship is highly dependent on the involvement of knowledgeable faculty members who can identify appropriate mentors and projects in industry. These faculty members should also be actively involved in supervision of the student's research in industry; ideally, the work should stem from a collaboration between a faculty member and an industrial scientist. Obviously, appropriate intellectual property agreements between the university and the company must be established to protect the freedom of the student ultimately to publish the data generated during the internship both in a dissertation and in scientific journals.

Through careful planning and coordination, approximately 50 students have completed industrial internships under the Pharmaceutical Biotechnology Program at the University of Kansas. All have been highly successful, particularly from the students' perspectives. The students not only have been able to do cutting-edge research in pharmaceutical biotechnology that has resulted in high-quality publications but also have gained exposure to the inner workings of biotechnology companies that has influenced their career choices. Faculty have also benefited, since the collaborations with industrial scientists that began through student internships have frequently continued long after the students' graduation.

In conclusion, the University of Kansas, through effective strategic planning and the support of an NIGMS predoctoral training grant, has been able to modify its graduate program in pharmaceutical chemistry to include aspects of biotechnology. This change was accomplished with the cooperation of the faculty from several academic departments, which allowed the creation of a multidisciplinary training program in pharmaceutical biotechnology. This program was structured in such a way that it did not detract from the graduate programs in the participating departments but, instead, helped to strengthen the departmental programs and make them more attractive to applicants.

The Future

To keep pace with ongoing changes in the pharmaceutical industry, the Department of Pharmaceutical Chemistry and the Pharmaceutical Biotechnology Program at the University of Kansas must continue to plan strategically and to evolve when necessary to continue to be competitive in selling their products (pharmaceutical scientists, knowledge, technologies) to their clients (pharmaceutical and biotechnology companies). The most significant recent changes include the introduction of new technologies into drug discovery (i.e., high-throughput screening, combinatorial chemistry, bioinformatics, genomics, proteomics), the integration of drug discovery and development, and the globalization of drug discovery and development. Through industrial internships, both students in the Department of Pharmaceutical Chemistry and trainees in the Pharmaceutical Biotechnology Program are being exposed to the technology revolution as well as to "integration initiatives" recently undertaken by the pharmaceutical and biotechnology industries. However, finding ways to expose our students to the globalization of these industries has been more difficult.

To increase the exposure of our graduate students to science at an international level and to facilitate increased interactions between our students and students and faculty from international universities, we established a not-for-profit organization, Globalization of Pharmaceutics Education Network, Inc. (GPEN, Inc.), in 1996. This organization now includes 25 universities—6 in the United States, 4 in Japan, 1 in Australia, and 14 in Europe. GPEN, Inc., sponsors a meeting every 2 years; the first was held at the University of Kansas in 1996, and the second took place at Galenische Pharmazie ETH in Zurich, Switzerland, in 1998. The third GPEN meeting will be held at the University of Uppsala in Sweden in 2000. These 3-day meetings focus largely on presentations by graduate students and discussions about graduate education in the participating universities (2 days). One day is set aside for short courses presented by the faculty on timely topics not necessarily covered in the curricula of the participating universities. Industry representatives are invited to take part, but their participation is limited to that of observers.

In the future, GPEN, Inc., plans to facilitate the exchange of students and faculty between participating universities. Such exchange programs would further globalize the education of graduate students in pharmaceutics, so that they are prepared to function in the international and emerging multinational pharmaceutical companies.

CONCLUSIONS

From my experiences as a mentor of graduate students, as a department chairperson, and as a director of a predoctoral training grant, I offer the following conclusions. While these conclusions appear to be specific to pharmaceutical chemistry, I believe that they are also generally applicable to most areas of chemistry.

- Changes in the pharmaceutical/biotechnology industry in the 1980s and 1990s have been rapid and dramatic, and additional changes are likely to occur in the future.
- The skill base required of an ideal pharmaceutical scientist in today's industry (Figure 14.5) is very different from that required of a scientist in the 1970s. The skill base required in 2010 will be very different from that required today.
- Pharmaceutical scientists must continue to be highly skilled in their areas of expertise in order to be productive and respected members of project teams.
- Pharmaceutical scientists today must be broadly trained so that they can communicate with scientists in other disciplines and work effectively on project teams.
- Training grants provide an excellent mechanism to facilitate change in academic programs and to build multidisciplinary training programs that can expand the scientific breadth of the students as well as enrich their scientific depth.

Small-Molecule-based Drugs	Macromolecule-based Drugs
Depth	**Depth**
• Organic chemistry	• Biochemistry
• Physical organic chemistry	• Organic chemistry
• Physical chemistry	• Biophysical chemistry
• Analytical chemistry	• Analytical chemistry
• Biopharmaceutics and pharmacokinetics	• Biopharmaceutics and pharmacokinetics
Breadth	**Breadth**
• Pharmacology	• Pharmacology
• Biochemistry	• Molecular biology
	• Bioengineering

FIGURE 14.5 Scientific depth and breadth needed by scientists who develop small molecule-based drugs or macromolecule-based drugs.

• Academic institutions have been slow to recognize the changes in the pharmaceutical industry and even slower to adjust their graduate programs to accommodate these changes.

The future strength of the scientific establishment in the United States is highly dependent on the willingness and the ability of academic units to change to accommodate the needs of their clients. Such change can arise only from effective strategic planning that focuses on the future and not the past.

RELATED READING

Ronald T. Borchardt, "Are Graduate Programs Training Pharmaceutical Scientists to Function Effectively in the New, Highly Integrated and Globalized Pharmaceutical Industry?" *Pharmaceutical Research* 14, 554-555, 1997.

Price Waterhouse Coopers, "Pharma 2005: An Industrial Revolution in R&D," Price Waterhouse Coopers, United Kingdom, 1998.

R. Hirschmann, "Introduction," pp. 1-5 in *Integration of Pharmaceutical Discovery and Development*, R.T. Borchardt, R.M. Freidinger, T.K. Sawyer, and P.L. Smith, eds., Plenum, New York, 1998.

DISCUSSION

Iwao Ojima, State University of New York at Stony Brook: I appreciate the kind of perspective that you brought and would like to get your insight about departments that are closely related to chemistry. In addition to the department of chemistry, you have at the University of Kansas a department of medicinal chemistry and pharmaceutical chemistry. So, I think you are in a position to observe and interact with those who are closely related. Can you comment on changes that have occurred in those two departments?

Ronald T. Borchardt: Let me begin by saying that we have a number of training grants at the University of Kansas, each in a different area. My presentation focused on the training grant in pharmaceutical biotechnology with the "core' department for this training grant being pharmaceutical chemistry. We also have a training grant on campus in the area of medicinal chemistry/pharmacology with the core departments being medicinal chemistry and pharmacology and toxicology. There is also a chemistry/biology interface training grant in which the core departments are chemistry and molecular biosciences.

Iwao Ojima: I am asking this question because I tried to hire a famous Russian physicist into our chemistry department, but I encountered a mind-set in the department againt the idea of hiring a physicist. I was shocked by this attitude. Do you know anything about the change in the chemistry department in Kansas?

Ronald T. Borchardt: One thing that I learned from being a department chair for 15 years was how to be somewhat diplomatic; therefore, I think it would be inappropriate for me to comment specifically about our chemistry department. However, I would say that our department, like many other chemistry departments, has gone through significant change in the past 20 years resulting from faculty retirements and the recruitment of young faculty members who are more receptive to multidisciplinary research and training programs.

P. Wyn Jennings, National Science Foundation: I want to comment on the multi-institutional idea that was raised previously for science and technology centers. I am not trying to evangelize with regard to our Integrative Graduate Education and Research Traineeship (IGERT) program, but I want to tell you that we do entertain multi-institutional proposals. Our experiences at this point—and we have several, somewhere between two and seven institutions involved in a single IGERT—is that an enlivened student is one who has studied for awhile in one institution, has gone away for 3 months to a year at another institution, and then has come back to the original institution. This individual is enlivened and infectious with enthusiasm. With regard to the international scene along this line, the IGERT program is negotiating a relationship with the megatraining centers in Germany and would like to be a participant likewise in programs elsewhere in the European Union. We are negotiating with them, and we are also involved with CNRS and the French. With regard to the need for foreign languages in different disciplines, I saw one experiment at the graduate student level in which three students from three different disciplines discussed the same topic in one seminar. It was extremely exciting.

Ronald T. Borchardt: With respect to industrial internships, our students have found them to be very positive experiences. They return from these industrial internships much more enthusiastic and focused on completing their Ph.D. degree. I think that this change in attitude arises because they realize what they are going to do when they graduate and are excited about the future.

Another comment about industrial internships is that the university must put in place an agreement with the company that completely protects the student. Whatever the student is doing in industry must be nonproprietary.

J. Michael White, University of Texas at Austin: I appreciate the care with which you stated the nature of these external collaborations. We need to be careful about just sending students off for a summer, because it can be extraordinarily counterproductive.

Ronald T. Borchardt: In our program, we carefully pick the company and scientist/mentor in the company so that we develop a true research collaboration for the student. Industrial scientists have many exciting ideas, but often they don't have time to pursue them. They are very excited about having the opportunity to mentor a graduate student.

James Nowick, University of California, Irvine: Let me start by commending you on the program that you have put together, as well as your presentation of it. Clearly, a lot of thought went into getting all of these areas together, supplementing the program when things were missing, and also responding to the needs of the industry. Irvine has a very zealous program of synthesis and natural-products-oriented synthesis. I would say that 90 percent or more of our students would have failed in being able to answer business-related questions. Yet, our students are gobbled up, particularly the ones from the straight synthesis background, by the pharmaceutical industry. It seems to me that there are multiple models for successfully training students. I would describe yours as more of a professional-school model; maybe the one for our synthetic organic chemists is more of an academic-department model, and they both seem to work. In our case, the students who come out and go on in the pharmaceutical industry quickly pick up the lingo and the issues about why the natural product they have synthesized isn't active. I would hate to think that our students, because of the fantastic depth they get, are wholly deficient in their training.

Ronald T. Borchardt: I find that graduates of chemistry departments are in general very, very bright, and they quickly learn what is important in the pharmaceutical industry. However, I must say that I was bothered by some of the comments I heard from the graduate students earlier in the workshop, i.e., that some chemistry mentors will not let their students out of the laboratory to attend seminars or courses in biochemistry or molecular biology. If that faculty member were in my department, there would be holy hell to be paid, because I think that students have to be given the opportunity, and even encouraged, when they are in the university to explore outside their major areas of study. To function effectively in the pharmaceutical industry, they need not only depth in their training but also breadth.

Robert L. Lichter, The Camille & Henry Dreyfus Foundation: Thank you for anticipating my question, but maybe you could elaborate. I have heard of a number of people who have traveled around exploring the notion of externships in one or another form, whether in industry or somewhere else, respond adversely to this idea. They see it as diverting students from the purposes for which they are in graduate school. I am wondering how either of you have dealt with it, to what extent have you run into it, and what the buy-in is for faculty—all of these kinds of ancillary issues.

Ronald T. Borchardt: When we developed the idea of internships in our training program, we decided that we were going to have our cake and eat it, too, in the sense that we are going to structure these experiences as research collaborations. As a result, these internships have ended up not extending the tenure of the students in our graduate program. In fact, I would say in some cases it may have expedited their graduate tenure, because they have gone to industry and done things that would have been difficult or impossible in the university.

Now, the other aspect of this, which I think is important, and which hasn't been brought up, is the fact that many of our internships have resulted in the establishment of faculty-scientist collaborations that have persisted long after the student has graduated. So, these student internships give the younger faculty the opportunity to get to know people in industry and better understand what their students will be doing when they graduate.

I have yet to see a serious downside to these internships. If you talked to the 50+ students in our department who have participated in internships in the past 10 years, they would all say that they had very positive experiences. Our faculty fully support internships. It turns out that in most cases I am the one who orchestrated the internships because I have the industrial contacts, but that is my responsibility as the program director of this training grant.

J. Michael White: You mentioned the word "contacts." I didn't discuss industrial interactions with our center, but what I think is valuable to remember is that those contacts are really important, and they need to be made at a variety of levels, especially in large organizations. There are opportunities for them at all kinds of places. You can get into Glaxo-Wellcome, Motorola, and so on, but you have to sustain those contacts over a long enough period of time so that these companies are willing to engage in projects with you.

Ronald T. Borchardt: I think the key is that the industrial contact must be at the scientist level. If you start with a manager or vice president, he or she can arrange an "industrial experience" for the student, but it will not necessarily be in the form of a research collaboration.

However, I must add that involvement of the managers and the vice president is important, because they are the people who can provide the funding. Funding for the student can come in various forms. Some pharmaceutical companies prefer to hire the student as a part-time employee. In this case, you must guard against the student's choice being driven by financial considerations. Sometimes companies give us unrestricted grants to supplement a student's stipend. This supplement is used to cover living expenses while the student is completing the internship.

Victor Vandell, Louisiana State University: I am a proponent of multidisciplinary training, and I am also a product of that, as my background is in organic and analytical chemistry. I think there is a resistance toward that type of approach. I would like to bring up an ongoing problem to show that multidisciplinary training is important. I have been hearing about the discounting of chemists in industry relative to chemical engineers. It bothers me that I could have a Ph.D. in chemistry and be placed on the same level, if not below the level, of a person with a bachelor's in chemical engineering. I perceive this as a future threat to chemists in industry. If we don't do something to start incorporating some type of overlap in our training of chemists so that they can take some chemical engineering courses or get more industrial experience to better compete, some industries might be leaning toward the ideology that they can actually replace us with chemical engineers. I want to know if anybody else perceives this as a threat.

Ronald T. Borchardt: I think that is a question that would be more appropriately directed to the chemistry faculty here than to me. I would make one real quick comment. As the pharmaceutical industry moves more to combinatorial chemistry and high-throughput screening, engineers become more and more important. Setting up robots to do the chemical syntheses and the biology testing is an engineering problem; thus, I would see an increasing need for graduates of engineering schools in the pharmaceutical industry.

Wayne Rohrbaugh, Ashland Chemical Company: If we can get out of the pharmaceutical focus for a moment, I think that there is a clear distinction between the contributions of a Ph.D. chemist and a chemical engineer in industry. Starting salaries, if that is what you are using as a benchmark, are primarily dictated by the law of supply and demand, and that is clearly the only thing that is going on there.

Participant: In other words, chemistry departments should go to the chemical engineering departments and talk about developing a course that takes students through the entire process of discovery through scale-up production. I think that would be of interest to chemical engineers as well as to chemists.

Robert Humphreys, National Starch and Chemical Company: I think that you would want to do that for the same reason that we would want to have a center, which is that people develop a common language. Chemical engineers and chemists do not speak a common language.

Wayne Rohrbaugh: I want to get away from the pharmaceutical focus a little bit because we have been looking at it so much. I have been sitting here for a day and a half, and I have been surprised at how introspective this group is. I have heard only tangential references to such things as the global economy, the impact of world market democratization, who your customers are, whether or not you have ever surveyed your customers and, by the way, your customers are industry, the chemical manufacturing industry in this country. According to the November 15, 1999, issue of *Chemical & Engineering News*, there are more than two times as many Ph.D. chemists in industry as in academia. If you look at full-time faculty and subtract out the effect of postdocs, it is closer to three times as many Ph.D.s in industry as there are in academia. So, the customer is missing.

Another thing I didn't hear anything about was the competitiveness of the U.S. chemical enterprise in the next 20 years. Are we going to be in existence in the next 20 years? What is the contribution of the chemical manufacturing industry to the gross domestic product of the United States? I can tell you it is quite substantial. I heard no description here whatsoever of the demographics of your placement of graduates as a function of industry sector. There is a very heavy emphasis on pharmaceutical research in this country. It is being supported by funding from the National Institutes of Health which is understandable, but the vast majority of National Science Foundation funding, and substantial amounts of Department of Energy funding are biotechnology related. And that accounts for 30 percent of the total research and development funding for the chemical enterprise in this country. The other 70 percent comes from industry. Clearly, the pharmaceutical industry is an important contributor to the gross domestic product, but it is only a fraction of the total chemical manufacturing contribution. I think you are neglecting the rest of us.

How many pharmaceutical biotechnology Ph.D.s will be needed in the next 10 to 20 years? Has that study been made? How many are you producing? What is happening with the law of supply and demand? Are those people going to be employable? Someone mentioned that he had to adjust his career many times; first he became a biochemist and then a pharmaceutical chemist. Was it by choice? How many of you in here started out as chemistry majors wanting to become biochemists? And how many are becoming biochemists? I think there is a fundamental bias right now for biotechnology, as induced from outside resources, the places to which you are going traditionally for funding. I am saying that the U.S. chemical industry is there also to provide funding. If we could break down the barriers of the university front office, we could work with you.

The majority of the U.S. chemical manufacturing industry, from my perspective at least, is not biotechnology related. It feels somewhat disenfranchised by the U.S. educational system, and we are becoming more enamored with European graduate schools. I want to make these comments and hear if we can get any discussion about them. Perhaps you should look at some of these data in the future in these kinds of discussions so that you can see where your product is going, not necessarily today, but in the next 20 years, and determine if you are prepared to supply that pipeline.

Ronald T. Borchardt: Let me take a minute to defend the content of my presentation. You must understand that I come from a department whose primary customer is the pharmaceutical industry. So, I think your question is probably more appropriately directed to the representatives of the chemistry departments in the audience.

Robert Humphreys: I want to pick up a little on what Ron said, particularly with respect to the idea of having a strategic plan. This forces you to develop a good model that other universities can look at, because the plan makes you look at who and where your customers are and, in this particular case, it happens to be pharmaceutical companies. You have to understand where the companies are going and what is driving their actions, in other words, understand your customers and in some cases maybe understand your customers' customers.

There are a lot of universities in this country and in the world competing with each other. National Starch spends a lot of money at universities outside the United States and for some very good reasons. This competition sets up a sort of war, just like that seen in industry, where the universities try to differentiate themselves. For example, if I look at Kansas, the first thing I think of is Dorothy from *The Wizard of Oz*. You are out there in the middle of the country and far from a pharmaceutical company. Yet, you have a need to differentiate yourself so when decisions are being made by a pharmaceutical company about whom to work for or work with and where its money should go, one thing that comes to mind right away is Kansas. Right?

Apparently you have been very successful based on the group of customers that you put up on the projector screen. You went through a strategic analysis and understood your customers in order to be successful in developing a plan and implementing it, and you differentiated yourself accordingly. This is the kind of model I think would be important for other universities that are thinking about how to attract and maintain a good base of industrial funding.

Another example of a university that has differentiated itself is Southern Mississippi, which went into the area of water-soluble polymers when everyone else was focusing on materials science. They have done a very good job of advertising it and attracting industrial funding for that reason. So, if this is the kind of thing you are interested in, I think that is a good model. My question would be what prompted you to go after them like that?

Ronald T. Borchardt: During my tenure as a department chair, I became a strong advocate of strategic planning. Our planning process started in 1984 when the faculty participated in a retreat. At that retreat, we asked ourselves, Who are our customers? What services are we providing to our customers? How are their worlds changing? This exercise was a very healthy experience. From this retreat, we developed a strategic plan that we now update annually.

Lynn Melton, University of Texas at Dallas: I am delighted by the compliments you are getting on your programs. I would like to remind people about the doctor of chemistry program at the University of Texas at Dallas, which has been in operation since 1983. We have had a 90 percent direct placement into industry and probably 60 industrial practicum students by now with 50 graduates. We asked our industrial friends what their jobs were like, and then we designed a curriculum to produce students who would be successful in the chemical process industry. We would be delighted to have any of you come on campus and find out whether our ideas are exportable to you. All students have an industrial internship of 9 to 12 months, paid for by industry, in which they are mentored as an industrial problem solver and solve problems to the benefit of the company. A lot of this material is available to you. We have taken it through the pilot stage and will be glad to share it.

Kathleen C. Taylor, General Motors: I think, Ron, you have managed to press the industrial hot button in this last session. I want to congratulate you on bringing up the globalization issue and the initiative that you have started. I think it is one of the most important things I have heard about at this meeting. I was interested to hear that foreign universities seem to be able to find a way to send students to work with you for these symposia. My question is, what have you learned in working with those institutions that is different, pro and con, that we should try to capture in our work in view of trying to bring globalization into the educational process?

Ronald T. Borchardt: I think the real question is what our students have learned, because this program is focused on providing new experiences for our graduate students.

Ernest L. Eliel, University of North Carolina: I am a schizophrenic in that I am both an elitist and an anti-elitist. Most of the time in these meetings, I put on my anti-elitist hat. I must now put on my elitist hat. In looking around the room, I have noticed that there are about a dozen universities in this country that have well-known chemistry departments that are not represented. I agree that one of the important functions of a university is to serve the student, and serving the student includes preparing him or her for a job. But there is another function of the university, which I have not heard about, and that is to advance the frontiers of knowledge. I think we must never forget that this is one of the purposes of the university and that we cannot be exclusively the handmaidens of industry. Universities have a divided role, and in our department, we have faced up to this divided role.

We have some faculty members who are collaborating with industries and even have their own businesses. Some also have very big interdisciplinary grants and turn out students in applied science fields who are readily hired. Others are doing synthesis, and their students are also readily hired. We also have some members on our faculty who are doing fundamental research whose impact cannot be seen perhaps for 10 or 20 years. If we stop doing fundamental research or research that does not yield an immediate return, I think we should close all of our universities and turn our attention to trade schools.

Ronald T. Borchardt: I fully agree with what you just said.

Closing Remarks

Robert L. Lichter, The Camille & Henry Dreyfus Foundation: Workshops have an interesting dynamic. No matter what their length, the last half of the last session is always the one at which people look at the clock. It's a natural response, even at this one where we are still on schedule.

First, I want to thank all of you for participating. As we said at the beginning, the success and vitality of the workshop would depend on your participation, a responsibility you fulfilled wonderfully. Indeed, your energy stood in marked contrast to the total silence I observed yesterday morning at the registration, a silence that we knew would not be sustained once the workshop was under way. You presented a variety of perspectives, which were discussed and debated energetically. There were a lot of sidebar conversations, always a sign of active engagement. The expressions of concern, even about the organization of the workshop itself, were forthright. Some of it was discomfiting, which is all to the good: we all learn from that experience. We talked about many issues and did not talk about others, such as the balance of professional and personal priorities, particularly for graduate students and young academic and industrial scientists. I can't help but comment on the notion of "customers" that was raised. I'm convinced that the number of definitions of "customer" maps directly onto the number of people who give the definition.

The question, of course, is What happens next? In the short term—about six months from now—a published proceedings will appear. In addition, we hope that you will examine how those thoughts, perspectives, and activities that were presented here and to which you have resonated may be realized in your own settings.

But the more difficult question is, What happens in the long term? A number of young scientists are here—graduate students, faculty members, and industrial scientists. Twenty years from now, will they be where we are, not merely looking at the same questions, but looking at them for the same reasons? As at least one speaker pointed out, education is a vital, vibrant, always changing enterprise. But what kind of change will take place and in what ways? Will the same kinds of questions be raised over and over again? Who will be the agents for change and under what circumstances?

Many questions remain open, which is what we hoped would be a result of the workshop. All of you did your jobs—presenters, participants, the organizing committee, and the National Research Council staff. I want to thank all of you for coming and wish you a safe voyage home.

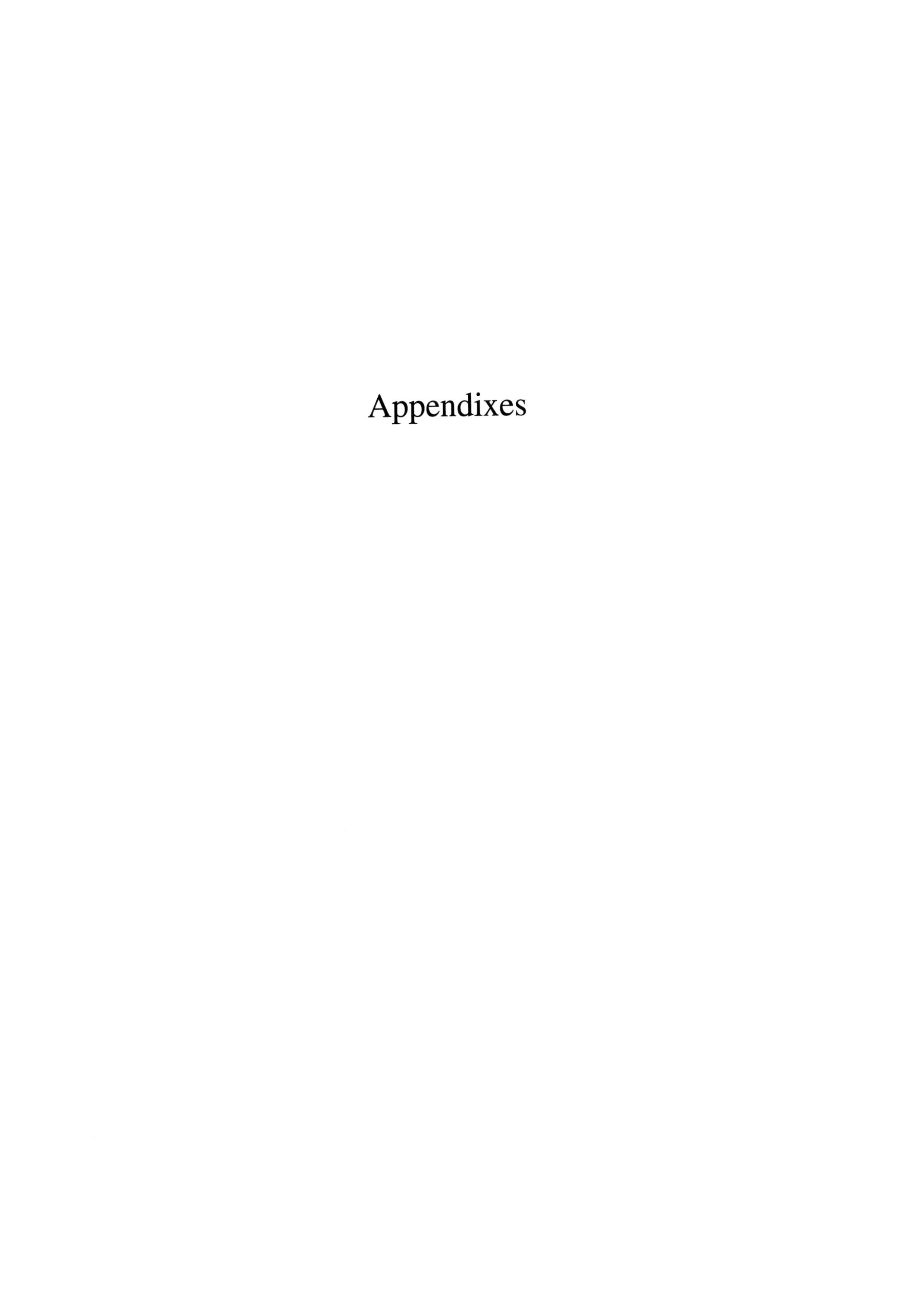

Appendixes

Appendix A

List of Workshop Participants

Joseph J.H. Ackerman, Washington University
Richard C. Alkire, University of Illinois at Urbana-Champaign
Larry B. Anderson, Ohio State University
Michael W. Babich, Florida Institute of Technology
Mark M. Banaszak-Holl, University of Michigan
Christopher F. Bauer, University of New Hampshire
Joseph J. BelBruno, Dartmouth College
Brian C. Benicewicz, Rensselaer Polytechnic Institute
David Bergbreiter, Texas A&M University
R. Stephen Berry, University of Chicago
Iona Black, Yale University
Ronald T. Borchardt, University of Kansas
Ronald Breslow, Columbia University
David E. Budil, Northeastern University
Jonathan L. Bundy, University of Maryland, College Park
Edwin A. Chandross, Bell Laboratories, Lucent Technologies
Steven S. Chuang, University of Akron
Robert E. Continetti, University of California, San Diego
Christopher J. Cramer, University of Minnesota
Glenn A. Crosby, Washington State University
Dady Dadyburjor, West Virginia University
Marcetta Y. Darensbourg, Texas A&M University
Peter K. Dorhout, Colorado State University
Michael P. Doyle, Research Corporation
Anne M. Duffy, University of California, San Diego
Thomas F. Edgar, University of Texas at Austin

Peter M. Eisenberger, Columbia University
Ernest L. Eliel, University of North Carolina
Billy Joe Evans, University of Michigan
Catherine Fenselau, University of Maryland
Ellen R. Fisher, Colorado State University
Marye Anne Fox, North Carolina State University
Joseph Francisco, Purdue University
James L. Fry, Colorado State University
Fred M. Hawkridge, Virginia Commonwealth University
Judson L. Haynes III, Procter & Gamble
Ned D. Heindel, Lehigh University
Michael J. Holland, Office of Management and Budget
Karlene A. Hoo, Texas Tech University
Robert W.R. Humphreys, National Starch and Chemical Company
John S. Hutchinson, Rice University
Madeleine Jacobs, American Chemical Society
Paul Jagodzinski, West Virginia University
Eric G. Jakobsson, University of Illinois at Urbana-Champaign
Lynn W. Jelinski, Louisiana State University
William S. Jenks, Iowa State University
P. Wyn Jennings, National Science Foundation
Donald E. Jones, National Science Foundation
Lynda M. Jordan, North Carolina Agricultural and Technical State University
Andrew Kaldor, Exxon Mobil Corporation
Timothy A. Keiderling, University of Illinois at Chicago
Melvin Koch, University of Washington
William F. Koch, National Institute of Standards and Technology
Brent Koplitz, Tulane University
Jay Labov, National Research Council
David K. Lavallee, State University of New York College at New Paltz
Abraham M. Lenhoff, University of Delaware
William A. Lester, Jr., University of California, Berkeley
Robert L. Lichter, The Camille & Henry Dreyfus Foundation, Inc.
Robert Lockhead, University of Southern Mississippi
David J. Malik, Indiana University-Purdue University Indianapolis
Robert S. Marianelli, Office of Science and Technology Policy
James D. Martin, North Carolina State University
Lynn A. Melton, University of Texas at Dallas
Craig A. Merlic, University of California, Los Angeles
François M.M. Morel, Princeton University
James S. Nowick, University of California, Irvine
Iwao Ojima, State University of New York at Stony Brook
James J. O'Malley, Exxon Chemical Company
Janet G. Osteryoung, National Science Foundation
David W. Oxtoby, University of Chicago
Soni O. Oyekan, Marathon Ashland Petroleum, LLC

Jeanne E. Pemberton, University of Arizona
Karen E.S. Phillips, Columbia University
Stanley H. Pine, California State University, Los Angeles
Lynmarie A. Posey, Michigan State University
C. Dale Poulter, University of Utah
Linda R. Raber, Chemical & Engineering News
Carolyn Ribes, The Dow Chemical Company
Charles Riordan, University of Delaware
Janet B. Robinson, University of Kansas
Michael E. Rogers, National Institutes of Health
Wayne J. Rohrbaugh, Ashland Chemical Company
Nina M. Roscher, American University
Stanley I. Sandler, University of Delaware
Barbara Sawrey, University of California, San Diego
John M. Schwab, National Institute of General Medical Sciences
Joel I. Shulman, Procter & Gamble
John P. (Jack) Simons, University of Utah
Jack Soloman, Praxair, Inc.
Nicholas H. Snow, Seton Hall University
Angelica M. Stacy, University of California, Berkeley
Derrick Tabor, National Institute of General Medical Sciences
Kathleen C. Taylor, General Motors
Ruthanne Thomas, University of North Texas
Laren M. Tolbert, Georgia Institute of Technology
Michael R. Topp, University of Pennsylvania
Victor Vandell, Louisiana State University
James W. Viers, Virginia Polytechnic Institute & State University
Sylvia Ware, American Chemical Society
Isiah M. Warner, Louisiana State University
John C. Warner, University of Massachusetts, Boston
Edel Wasserman, E.I. du Pont de Nemours & Company
Richard A. Weibl, Association of American Colleges and Universities
Jodi L. Wesemann, St. Mary's College of California
J. Michael White, University of Texas at Austin
Frankie K. Wood-Black, Phillips Petroleum
Bettina J. Woodford, University of Washington
John T. Yates, Jr., University of Pittsburgh

Staff
Maria P. Jones
Ruth McDiarmid
Sybil A. Paige

Appendix B

Biographical Sketches of Workshop Speakers

R. Stephen Berry is the James Franck Distinguished Service Professor of Chemistry at the University of Chicago. He has worked on a variety of subjects ranging from strictly scientific matters to the analysis of energy use and resource policy. His scientific research has been in part theoretical, in areas of finite-time thermodynamics, atomic collisions, atomic and molecular clusters and chaos, and in part experimental, involving studies of chemical reactions and laser-matter interactions. Some of his work outside traditional science has involved interweaving thermodynamics with economics and resource policy. He has also worked since the mid-1970s with issues of science and the law, and with management of scientific data, activities that have brought him into the arena of electronic media for scientific information and issues of intellectual property in that context. His current interests include the dynamics of atomic and molecular clusters, the thermodynamics of time-constrained processes and the efficient use of energy, and a variety of issues concerning science and public policy, including precollegiate education and scientific literacy, the maintenance of scientific enterprises in America and elsewhere, the impact of electronic communication on the sciences, and the conduct of scholarly work in general. Dr. Berry received his A.B., A.M., and Ph.D. in 1952, 1954, and 1956, respectively, from Harvard University. He is a member of the National Academy of Sciences.

Ronald T. Borchardt is the Solon E. Summerfield Distinguished Professor of Pharmaceutical Chemistry at the University of Kansas-Lawrence. He received his undergraduate education (B.S. in pharmacy, 1967) from the University of Wisconsin-Madison and his graduate education (Ph.D. in medicinal chemistry, 1970) from the University of Kansas-Lawrence. After serving as a postdoctoral fellow at the National Institutes of Health from 1969 to 1971, Professor Borchardt returned to the University of Kansas as an assistant professor in the Department of Biochemistry in the College of Liberal Arts and Sciences. In the 1970s he was promoted through the academic ranks to his current position. From 1983 to 1998, he served as the chair of the Department of Pharmaceutical Chemistry in the School of Pharmacy. During his academic career Professor Borchardt has received numerous awards and honors for his teaching and research accomplishments and is the author or co-author of approximately

425 scientific publications and 375 abstracts. He is also the editor of six books and the series editor of *Pharmaceutical Biotechnology*. His research interests are focused in the areas of drug design and drug delivery.

Jonathan L. Bundy graduated from North Carolina State University with a B.S. degree in biochemistry and did undergraduate research in the laboratory of Dr. Jim Otvos. Later that year he began graduate study in the biomedical sciences program at Hood College, doing research in biological mass spectrometry with Harry Hines of the U.S. Army Medical Research Institute of Infectious Diseases. In 1997, he transferred to the doctoral program at the University of Maryland, Baltimore County, and began research under the direction of Catherine Fenselau. Mr. Bundy is currently completing his Ph.D. studies with Dr. Fenselau at the University of Maryland, College Park, where she moved in 1998 to become chair of its Department of Chemistry and Biochemistry. His research interests are centered on the development of biomolecule-derivatized surfaces for mass spectrometric analysis of microorganisms.

Peter M. Eisenberger is a professor of earth and environmental sciences at Columbia University. He received a B.A. in physics with honors from Princeton University in 1963 and graduated in 1967 with a Ph.D. in applied physics from Harvard University, where he received both a Woodrow Wilson and a Harvard Fellowship and remained for one year as a postdoctoral fellow researching both biophysics and the polaron problem. In 1968 he joined Bell Laboratories and held the post of department head from 1974 to 1981. In 1981, he joined Exxon Research and Engineering Company as director of its Physical Sciences Laboratory and was appointed senior director in charge of Exxon's Corporate Research Laboratory in 1984. In 1989, he was appointed professor of physics and director of the Princeton Materials Institute at Princeton University. From 1996 to 1999 he held the posts of vice provost of the Earth Institute and director of the Lamont-Doherty Earth Observatory.

Marye Anne Fox is chancellor of North Carolina State University in Raleigh, North Carolina. Prior to assuming her current post in 1999, she was vice president for research and the M. June and J. Virgil Waggoner Regents Chair in Chemistry at the University of Texas at Austin. Her recent research activities include organic photochemistry, electrochemistry, and physical organic mechanisms. She is a former associate editor of the *Journal of the American Chemical Society*. Previously, she was director of the Center for Fast Kinetics Research, vice chair of the National Science Board, and a member of the Task Force on Alternative Futures for the Department of Energy National Laboratories (the Galvin Committee). Dr. Fox is a member of the National Academy of Sciences (NAS), has served on the NAS Council Executive Committee, and is a member of the NAS Committee on Science, Engineering, and Public Policy. She is a former member of the National Research Council's Commission on Physical Sciences, Mathematics, and Applications and served on the Committee on Criteria for Federal Support of Research and Development. She received her bachelor's degree from Notre Dame College, a master's degree from Cleveland State University, and her Ph.D. in organic chemistry from Dartmouth College.

Judson L. Haynes III received a bachelor of arts degree in chemistry from Hampton University, where he was a Minority Access to Research Careers (MARC) Scholar. Dr. Haynes went on to enter the graduate program as a National Institutes of Health Predoctoral Fellow in the Department of Chemistry at Louisiana State University (LSU) in Baton Rouge, Louisiana. While at LSU, he worked in the area of capillary electrophoresis under the direction of Dr. Isiah M. Warner. Specifically, he has developed novel pseudo-stationary phases (such as dendrimers, cyclodextrins, and micelle polymers) for separations in electrokinetic chromatography. Currently, Dr. Haynes is a research scientist in the Baby Care Analytical Section at the Procter & Gamble Company.

Eric G. Jakobsson was trained as a chemical engineer (B.S. 1960, Columbia) and a physicist (Ph.D., 1969, Dartmouth). He became interested in biology through an interest in electrical excitability in nerves, and after a postdoctoral stint in the Department of Physiology at Case Western Reserve came to the University of Illinois Department of Physiology and Biophysics in 1971, where he has spent his career. At different times he has worked on gating of ion channels, permeation through ion channels, functional organization of epithelia, molecular structure of membranes, metabolism, and education research. For his work in ion permeation, he was elected a fellow of the American Physical Society in 1994. Dr. Jakobsson has been affiliated with the National Center for Supercomputing Applications since 1991. He was a member of the team that invented and developed the Biology Workbench and currently heads a National Science Foundation-sponsored project that develops educational applications of the Workbench.

Lynn W. Jelinski is vice chancellor for research and graduate studies and dean of the graduate school at Louisiana State University, where she is responsible for the university's research programs, research centers, graduate school, technology transfer, and economic development. From 1991 to 1998, Dr. Jelinski served as director for the Cornell Center for Advanced Technology (Biotechnology) and from 1997 to 1998 served also as director of the Cornell Office of Economic Development. She previously headed the Biophysics Research and Polymer Chemistry Research Departments at AT&T Bell Laboratories. She received her doctorate in chemistry at the University of Hawaii in 1976. Her research interests include solid state nuclear magnetic resonance spectroscopy and its application to elucidate the molecular mechanism for the strength of spider silk. Dr. Jelinski has over 100 refereed publications in journals such as *Science, Nature, Physical Review Letters, Macromolecules,* and the *Journal of the American Chemical Society.*

François M.M. Morel is Albert G. Blanke Professor of Geosciences at Princeton University. He is also director of the Princeton Environmental Institute and Director of the Center for Environmental Bio-inorganic Chemistry in Princeton, New Jersey. He is a visiting professor at the Université de Paris VI. His major fields of interest are aquatic chemistry and aquatic biology with a focus on the interactions of trace elements and microbiota and the role of trace elements in the global carbon cycle. Before joining Princeton University he was for 20 years a professor at the Massachusetts Institute of Technology, where he served as director of the R.M. Parsons Laboratory. He has served on many national and international committees dealing with environmental issues.

Karen E.S. Phillips earned an associate's degree in chemistry with a minor in fine art from Miami-Dade Community College in Florida. She completed her undergraduate degree at Barry University, also in Florida, with a major in chemistry and minors in both biology and fine art. At Barry University, she was a research fellow in the Minority Access to Research Careers (MARC) program, working on the synthesis of muscarinic agonists for Alzheimer's therapy. Ms. Phillips entered the Ph.D. program at Columbia University after completing her undergraduate degree and became a founding member of the Columbia Chemistry Careers Committee. She is currently completing her Ph.D. work on the synthesis of aggregating heterocyclic helicenes with Dr. Thomas Katz.

Angelica M. Stacy received her B.A. degree from LaSalle College in 1977 and her Ph.D. from Cornell University in 1981 with Professor M.J. Sienko, and then went on to do postdoctoral research with professors R.P. van Duyne and P. Stair at Northwestern University. She joined the faculty in the Chemistry Department of the University of California, Berkeley, in 1983, where she is now a full professor. Her research interests are in the areas of materials chemistry and chemistry education.

Edel Wasserman obtained a B.A. in chemistry from Cornell University in 1953 and an M.A. and a Ph.D. at Harvard University under professors William E. Moffitt and Paul D. Bartlett. He joined Bell Laboratories in 1957. Beginning in 1967 he held joint appointments as a member of the technical staff at Bell Laboratories and as professor of chemistry at Rutgers University. He joined Allied Chemical Corporation in 1976 as director of chemical research and later became director of corporate research. He moved to Central Research & Development at DuPont in 1981, where he is now science advisor. He served as president of the American Chemical Society in 1999.

Richard A. Weibl is director of programs in the office of education and institutions renewal at the Association of American Colleges and Universities (AAC&U). He came to AAC&U specifically to give leadership to the Preparing Future Faculty (PFF) program. This highly collaborative national program is designed to change the educational experiences of future faculty. The program has grown from 15 clusters of institutional partners to include 20 science and mathematics departmentally based clusters and will soon add 24 social science and humanities clusters. In all, more than 200 schools are participating in funded PFF programs, and dozens of others have created programs based on the PFF model. Dr. Weibl came to AAC&U from Antioch College, where he served as director of Institutional Research and Evaluation Studies. Prior to Antioch, he did doctoral studies at the Ohio State University in educational policy and leadership and worked in student affairs at Longwood College and Marquette University. He holds a master's degree from the University of Georgia and a bachelor's degree from Bowling Green State University.

J. Michael White received his bachelor of science in chemistry from Harding College in 1960 and his Ph.D. in chemistry from the University of Illinois in 1966, and then joined the chemistry faculty at the University of Texas at Austin as assistant professor. He was named associate professor in 1970 and full professor in 1976. From 1979 to 1984, he served as chair of the department, and he has held the Norman Hackerman Professorship in Chemistry since 1985. Since 1991, he has directed the National Science Foundation (NSF)-supported Science and Technology Center for Synthesis, Growth, and Analysis of Electronic Materials at the University of Texas. Since 1976, he has been a visiting staff member at Los Alamos National Laboratory. Dr. White served as a program officer at the NSF in 1978-1979 and was a summer guest worker at the National Bureau of Standards during the same period. He and his students have enjoyed long-term interactions with Sandia National Laboratories. He is actively working on problems in surface chemistry, the dynamics of surface reactions, and photo-assisted surface reactions.

Appendix C

Origin of and Information on the Chemical Sciences Roundtable

In April 1994, the American Chemical Society (ACS) held an Interactive Presidential Colloquium entitled "Shaping the Future: The Chemical Research Environment in the Next Century."[1] The report from this colloquium identified several objectives, including the need to ensure communication on key issues among government, industry, and university representatives. The rapidly changing environment in the United States for science and technology has created a number of stresses on the chemical enterprise. The stresses are particularly important with regard to the chemical industry, which is a major segment of U.S. industry, makes a strong, positive contribution to the U.S. balance of trade, and provides major employment opportunities for a technical work force. A neutral and credible forum for communication among all segments of the enterprise could enhance the future well-being of chemical science and technology.

After the report was issued, a formal request for such a roundtable activity was transmitted to Dr. Bruce M. Alberts, chairman of the National Research Council (NRC), by the Federal Interagency Chemistry Representatives, an informal organization of representatives from the various federal agencies that support chemical research. As part of the NRC, the Board on Chemical Sciences and Technology (BCST) can provide an intellectual focus on issues and fundamentals of science and technology across the broad fields of chemistry and chemical engineering. In the winter of 1996, Dr. Alberts asked BCST to establish the Chemical Sciences Roundtable to provide a mechanism for initiating and maintaining the dialogue envisioned in the ACS report.

The mission of the Chemical Sciences Roundtable is to provide a science-oriented, apolitical forum to enhance understanding of the critical issues in chemical science and technology affecting the government, industrial, and academic sectors. To support this mission, the Chemical Sciences Roundtable will do the following:

[1] *Shaping the Future: The Chemical Research Environment in the Next Century,* American Chemical Society Report from the Interactive Presidential Colloquium, April 7-9, 1994, Washington, D.C.

- Identify topics of importance to the chemical science and technology community by holding periodic discussions and presentations, and gathering input from the broadest possible set of constituencies involved in chemical science and technology.
- Organize workshops and symposia and publish reports on topics important to the continuing health and advancement of chemical science and technology.
- Disseminate the information and knowledge gained in the workshops and reports to the chemical science and technology community through discussions with, presentations to, and engagement of other forums and organizations.
- Bring topics deserving further, in-depth study to the attention of the NRC's Board on Chemical Sciences and Technology. The roundtable itself will not attempt to resolve the issues and problems that it identifies—it will make no recommendations, nor provide any specific guidance. Rather, the goal of the roundtable is to ensure a full and meaningful discussion of the identified topics so that the participants in the workshops and the community as a whole can determine the best courses of action.